AF183696

SpringerBriefs in Geography

SpringerBriefs in Geography presents concise summaries of cutting-edge research and practical applications across the fields of physical, environmental and human geography. It publishes compact refereed monographs under the editorial supervision of an international advisory board with the aim to publish 8 to 12 weeks after acceptance. Volumes are compact, 50 to 125 pages, with a clear focus. The series covers a range of content from professional to academic such as: timely reports of state-of-the art analytical techniques, bridges between new research results, snapshots of hot and/or emerging topics, elaborated thesis, literature reviews, and in-depth case studies.

The scope of the series spans the entire field of geography, with a view to significantly advance research. The character of the series is international and multidisciplinary and will include research areas such as: GIS/cartography, remote sensing, geographical education, geospatial analysis, techniques and modeling, landscape/regional and urban planning, economic geography, housing and the built environment, and quantitative geography. Volumes in this series may analyze past, present and/or future trends, as well as their determinants and consequences. Both solicited and unsolicited manuscripts are considered for publication in this series.

SpringerBriefs in Geography will be of interest to a wide range of individuals with interests in physical, environmental and human geography as well as for researchers from allied disciplines.

More information about this series at http://www.springer.com/series/10050

Julian Bolleter · Cristina E. Ramalho

Greenspace-Oriented Development

Reconciling Urban Density and Nature in Suburban Cities

Springer

Julian Bolleter
Australian Urban Design Research Centre (AUDRC)
The University of Western Australia
Perth, WA, Australia

Cristina E. Ramalho
School of Biological Sciences, NESP Clean Air and Urban Landscapes Hub
The University of Western Australia
Perth, WA, Australia

ISSN 2211-4165 ISSN 2211-4173 (electronic)
SpringerBriefs in Geography
ISBN 978-3-030-29600-1 ISBN 978-3-030-29601-8 (eBook)
https://doi.org/10.1007/978-3-030-29601-8

This Springer imprint is published by the registered company Springer Nature Switzerland AG
The registered company address is: Gewerbestrasse 11, 6330 Cham, Switzerland

Foreword

OMG: Look at all the people in the park!

In the western suburbs of Sydney in the 60s and 70s kids ran in feral packs. We never wore shoes or used sunscreen. Bottled water hadn't yet been invented, pools were uncommon luxury items and the idea of carrying a phone around with you would have been absurd. Parks were nothing but big flat grassed areas with maybe a swing off to one side, and a concrete cricket pitch in the middle.

Instead, the whole suburb was an adventure playground. From dawn to dusk, we would be free-ranging through the neighbourhood's massive back yards: climbing trees, pinching fruit, rifling through sheds, careening down wide streets with crusty "nature strips" on cheap dragsters, or pretty much anything with wheels. Equipped with essentials like compasses and sling-shots, we'd fan out across housing estates under construction, rummaging around the building sites, chucking rocks, investigating derelict farms and hunting for wildlife in what seemed like vast tracts of pristine bush and open creeks.

But then stacks of storm water pipes and surveyor's pegs started appearing. Baby Kookaburras and the odd owl would mysteriously just show up in the garden, dazed. We'd feed them minced meat or whatever, and if that didn't kill them then the cats and dogs did. I once clubbed a red belly black snake to death and took it to school as a trophy. Little did we know, this endless playscape was about to vanish under the tide of Sydney's manifest westward destiny. Indeed, we were that tide.

However, as a teenager in the 80s, the tide turned and we all headed in the opposite direction, making a beeline for that glowing thing called "the city". This meant the inner city—Newtown, Surry Hills, Darlinghurst, etc.; a promiscuous labyrinth of cheap rental space and "leb" food. Terraces and warehouses overflowed with undernourished mods, goths, punks and vestigial "westies" not yet prepared to completely reinvent themselves. On every corner was a pub, and in every pub was a band. Everyone was an artist, or could at least look like one. A ban on al-fresco dining finally gave way and "café culture" spilled into the streets. Sydney became the Gay centre of the universe, and the miracle of the city's harborside parks was always there to help nurse hangovers the next day.

So why introduce a book about increasing density around public open space in Australian suburbs with reminiscences of low-density suburbia on the one hand, and high-density urbanity on the other? Well, because with this book landscape architect, Julian Bolleter and urban ecologist, Cristina Ramalho are suggesting that, within reason, these two forms of urbanism are not contradictory; they are not just of the past, and most importantly they don't necessarily need to be estranged from one another at separate ends of the city. They argue that with a little faith in GOD: not the deity, but a new urban development approach, Greenspace-Oriented Development, where medium- to high-density development is grafted tightly onto existing public open spaces, we can bring together the virtues of suburbia and the inner city.

This is not the first time planners have tried to mix landscape and urbanity in a bid to offer the best of both. But whereas previous models such as Le Corbusier's "Towers in the Park", Ebenezer Howard's "Garden City" and Frank Lloyd Wright's "Broadacre City" demanded nothing less than the complete restructuring of the existing city, and failed as a consequence, Bolleter and Ramalho urge that we act now with relative precision. They show how we can both enhance and preserve the existing urban fabric of our cities whilst increasing density around the edges of parks and other tracts of open space.

Doing so not only offers public open space in lieu of private backyards to incentivize healthy apartment living, but it could also obviate the current practice of low-quality urban infill, such as the battle axing of middle-ring suburban blocks. This would help save backyards for the kids of the future, give the urban forest the space it requires to mature, and also ensure porous land to absorb and filter the storm water run-off that would otherwise pollute our waterways. Not only that, more people living directly adjacent to parks translates into more activity in, and money for, said parks, many of which are currently underutilized burdens on local government budgets.

Densification around public open spaces would also make our parks more social. Parks could become to Australians what piazzas are to Italians. More people living closer to parks could also bring people and the natural world closer together. Our parks could be retrofitted as oases of biodiversity and providers of other ecosystem services. Most importantly, from an environmental perspective, increasing the density of the existing urban fabric will help mitigate sprawl at the edge of the city, where species are not just being killed by wild kids, but by bulldozers preparing new suburbs (without big backyards).

Despite the obvious merits of GOD, I can already hear the nay-sayers lining up: the land around parks won't be cheap to parcel, the NIMBYs will be apoplectic and finally the big no-no: GOD is not TOD so it reinforces a disconnect between density and public transport. We should note here that Bolleter and Ramalho do draw a line around where GOD should take place, suggesting that sites within a 20-minute walkable catchment around public transport nodes are best. But even if these are legitimate concerns, do they outweigh GODs' possible benefits? In shooting down new ideas we should also note that GOD is not being presented here as a cure-all. Australian cities need a mosaic of solutions for their growing populations; solutions that capitalize on our cities' existing assets, that preserve and enhance livability and that offer a range of lifestyle options, at a range of price points. GOD doesn't replace TOD, but I for one would rather live next to the park than the train station. I loathe the way Australian planning is reducing the city to a question of getting to and from work.

Comparing GOD to my own nostalgia for Australian cities, I can imagine that being a twenty-first-century kid in an apartment close to a park would be pretty good. I can also imagine being a young adult in a suburb with at least some enclaves of density to spice things up a bit, and the park as rendezvous also being OK. And looking to the future I can also imagine downsizing and retiring with a park view and a café on the corner.

God knows, you could do a lot worse.

Richard Weller
Professor and Chair of Landscape Architecture
Meyerson Chair of Urbanism
Co-Director, McHarg Center
Department of Landscape Architecture
Stuart Weitzman School of Design
University of Pennsylvania
Philadelphia, PA, USA

Acknowledgements

Acknowledgements for Julian Bolleter

Thank you to the supporters of the Australian Urban Design Research Centre where I work, The Western Australian Planning Commission, The Western Australian Department of Planning, Lands and Heritage, the Department of Communities and Landcorp.

Thank you to the Journal of Landscape Architecture and Australian Planner who have kindly granted permission for us to reproduce, with edits, material in this book. The original papers are:

Bolleter, J. (2016). Background noise: a review of the effects of background infill on urban liveability in Perth. Australian Planner, 10, 1–14.

Bolleter, Julian, & Ramalho, Cristina E. (2014). The potential of ecologically enhanced urban parks to encourage and catalyze densification in greyfield suburbs. Journal of Landscape Architecture, 9(3), 54–65.

Thanks to Rob Cameron who assisted with the graphic production, and to Paula Hooper and Bill Grace for their astute reviews.

I am grateful to my colleagues, Anthony Duckworth-Smith, Bill Grace, Grace Oliver, Jill Penter, Paula Hooper and Zoe Myers for their tolerance of yet another book project.

I would like to acknowledge Richard Weller, who in 2012 introduced me to the concept of correlating urban density and upgraded green space—a concept which forms the basis of this book.

Thank you also to my co-author Cristina Ramalho for a productive and pleasurable collaboration.

Thanks go to my parents, Glenys and Ross, for their unwavering support.

Finally, I am grateful to my partner, Dr. Sally Appleton, and my daughter Rose Bolleter, for their love and support.

Acknowledgements for Cristina E. Ramalho

I would like to acknowledge the support by the Australian Government's National Environmental Science Program through the Clean Air and Urban Landscapes Hub.

I would also like to thank my colleagues within the Hub, as well as those within the Richard Hobbs' Ecosystem Restoration and Intervention Ecology group, for their general support, stimulating work environment and academic discussions. A special thanks to Caragh Threlfall, Kylie Soanes, Natasha Pauli, Kate Lee, Dawn Dickinson, Leonie Valentine, David Kendal and Judy Bush for the discussions, shared learning and collaborative work on the topic of urban green spaces and urban biodiversity.

Thank you also to Julian Bolleter for the ongoing much valued collaboration and the opportunity to co-write this book.

Finally, to my partner Jens Kirsch and our daughter Isabelle who was born just before this book was submitted for publication, to you both, thank you for your patience, inspiration and love.

Contents

About the Authors

Dr. Julian Bolleter is Co-Director of the Australian Urban Design Research Centre (AUDRC) at The University of Western Australia. His role at AUDRC includes teaching a master's program in urban design and conducting urban design-related research and design projects. He is a landscape architect and urban designer and has worked in Australia, the USA, the UK and the Middle East. He has completed a Ph.D. concerning urban development in Dubai and has published six books. His research focusses on the design of new cities, urban densification, and multifunctional public open space. He has received funding from the Australian Research Council, Healthways and the Western Australian government.

Dr. Cristina E. Ramalho is a Research Fellow in Urban Ecology at The University of Western Australia. She is Leader of the *Urban Greening for Livability and Biodiversity* Project within the Clean Air and Urban Landscapes Hub of the Australian National Environmental Science Program. Her work is inter-disciplinary and focuses on how we can better plan, design and manage urban environments in order to make these more livable and biodiverse. She is particularly interested in (1) understanding how the design of urban green spaces can be better informed by multidisciplinary knowledge aiming to optimize their socio-ecological benefits; (2) conservation of urban biodiversity, especially remnant plant communities; and (3) integration of traditional knowledge in land-use and water planning and biodiversity conservation.

Abbreviations

GOD	Greenspace-Oriented Development
ha	hectare(s)
km	kilometer(s)
m	metre
POS	Public Open Space
TOD	Transit-Oriented Development

Chapter 1
Introduction

Abstract This introductory chapter establishes the background within which the book's discussion about urban densification is set. Urban sprawl is a major facet of contemporary urbanization. In countries such as Australia and the US, expansive suburbanization consisting of single-family, freestanding residential dwellings is ubiquitous. Despite its ubiquity, urban sprawl faces several emerging crises, which threaten its dominance. These include the destruction of agriculturally productive and biodiverse land, ballooning infrastructure costs and commuting times, and the concentration of vulnerable socio-economic strata on the city's fringes. These issues, in conjunction with the challenges of delivering Transit-Oriented Development in existing urban areas, highlight the need for a complementary strategy for achieving urban densification in suburban cities. In this book we explore this topic with a focus on the Australian context.

Keywords Cities · Compact cities · Greenspace-Oriented Development · Transit-Oriented Development · Suburbs · Urban infill · Urban consolidation · Urban densification · Urban green space · Urban parks · Urban sprawl · Public open space

1.1 Background

1.1.1 Global Urban Growth

We live on an urbanizing planet. In 2008, demographers at the United Nations announced that more than 50% of humans were living in cities and the sprawling suburbs that surround them. This symbolized a profound change in human history. Never has most of the world's population lived in urban areas. The world today has 4.2 billion urban dwellers (United Nations 2018), 500 cities with populations of at least one million people, 74 with at least five million, and 12 with at least 20 million (McNeill and Engelke 2016). By 2050, there will be 2.5 billion more urban dwellers, accounting for 68% of the world's population (United Nations 2018), and making urbanization one of the twenty-first-century's most transformative trends (United Nations General Assembly 2016).

J. Bolleter and C. E. Ramalho, *Greenspace-Oriented Development*,
SpringerBriefs in Geography, https://doi.org/10.1007/978-3-030-29601-8_1

Due to rapid urbanization, cities will continue to be the focus for population growth, economic activity, social, cultural and environmental issues (United Nations General Assembly 2016). Cities are already, and increasingly will be, confronted with enormous challenges in terms of the provision of housing, infrastructure, basic services, health, education, employment and natural resources, amongst many others (United Nations General Assembly 2016). Clearly, the planning and design of our cities will fundamentally determine the viability of the human species—amongst many others—in this twenty-first century.

1.1.2 Mass Suburbanization

Around the world, the vast majority of people are flocking to cities not to dwell in their centres but to suburbanize their peripheries (Berger et al. 2017). Reflecting this, from 2000 to 2015, in all regions of the world, the expansion of urbanized land outpaced the growth of urban populations, resulting in unprecedented "urban sprawl" (United Nations 2017) (Fig. 1.1). As a result of population growth, and in some cases declining densities, by 2030, an estimated 1.2 million km^2 of land worldwide will become urbanized (Seto et al. 2012).

Fig. 1.1 Suburbia: From 2000 to 2015, in all regions of the world, the expansion of urbanized land outpaced the growth of urban populations, resulting in unprecedented "urban sprawl". *Source* Julian Bolleter

In the western world,[1] as soon as the middle class could afford to move to low-density suburbia, development at the periphery has boomed (Bruegmann 2017). As a result, in countries such as Australia, the US and UK, expansive suburbanization consisting of single-family, freestanding housing is ubiquitous (e.g. Schneider and Woodcock 2008). Indeed, over 80% of the present-day populations in the US and UK find the suburbs "attractive places to live" (Hagan 2017).

There is a broad consensus in the literature that most Australians also aspire to own a large, detached house in the suburbs (Kelly et al. 2011a) (Fig. 1.2). The "Australian dream" of owning your own home is often automatically associated with a detached house on a block of land and is seen as a mark of having "made it". For instance, a study conducted in Perth, revealed that when not constrained by income, 79% of people preferred a separate dwelling and 13% a semi-detached option, with only 7% preferring flats, units or apartments (Curtin University and Hames Sharley 2013).[2] Evidently, the suburban dream runs deep in the Australian cultural psyche (Kelly et al. 2011a), and Australia is not alone. As a result of widespread preference for

Fig. 1.2 The "suburban dream": Most Australians aspire to own a large, detached house in the suburbs. This is in part because suburban gardens offer the opportunity to grow food, have pets, entertain and relax in private and in nature. *Source* Image by Rennie Ellis courtesy of the National Library of Australia (https://trove.nla.gov.au/version/192344130)

[1]By western world, we mean principally Australasia, North America and Western Europe.

[2]However, when researchers asked respondents what home they would like to live in, taking into account realities such as housing costs and income, respondents indicated that compromises such as semi-detached dwellings or, in some cases, apartments were satisfactory (Kelly et al. 2011b).

suburban living, the Economist declared, in a recent essay, that globally we are not in the age of urbanization but rather the age of suburbanization (Wall 2017).

1.1.3 The Problems of Urban Sprawl

Despite its enduring popularity, suburban sprawl faces several emerging crises, which threaten its dominance. These include challenges in relation to the destruction of agriculturally productive (e.g. Seto et al. 2000) and biodiverse land in the peri-urban zones (e.g. Radeloff et al. 2010), ballooning service and public transport infrastructure costs (Brownstone and Golob 2009; Bento et al. 2005), and the concentration of socio-economic vulnerabilities on the city's fringes (Dodson and Sipe 2008; Zhao and Kaestner 2010; Sturm and Cohen 2004; Nechyba and Walsh 2004) (Fig. 1.3).

Life for residents of the suburban periphery can be dominated by longer commutes, which researchers have linked to lower overall well-being and life satisfaction (Kelly et al. 2012). Indeed, more than a quarter of all commuters in Australia's big cities spend more time commuting than they do with their children (Kelly and Donegan 2015). Outer suburbs also offer poorer access to jobs, which affects residents' ability to maintain and develop a career (Kelly and Donegan 2015). Concomitantly, newer outer suburbs often offer significantly less private garden space than their middle-ring

Fig. 1.3 The problems of sprawl: Contractors clear once biodiverse land on the edge of Perth for suburban sprawl. *Source* Donna Broun, Richard Weller

equivalents (Bolleter 2017) due to New Urbanism-led attempts to deliver "compact" suburbs. Such approach can result in outer suburbs that do not offer the "leafy green" qualities of long-established suburbs nor the true benefits of urbanity, such as walkability, vibrancy and diversity.

1.1.4 Planning for Urban Densification

To alleviate the impacts of urban sprawl, mitigate and adapt to climate change, improve human well-being, and advance societal inclusivity, there is an emphasis on delivering urban densification in most cities worldwide. In pursuit of urban infill development, urban planning strategies have focused primarily on Transit-Oriented Development (TOD) principles (Calthorpe and Fulton 2001), which advocate densification around public transport nodes and corridors (City of Melbourne 2010; Woodcock et al. 2010; Dittmar et al. 2004).

Despite the widespread adoption of TOD ideology, many urban planning agencies are not achieving their targets for urban infill (Bolleter and Weller 2013; Berger et al. 2017). In part, this is because the TOD planning strategies follow a rather "inflexible, over-neat vision" of cities (…) that sits at odds with their "increasing geographical complexity" (Forster 2006). Moreover, much of the constructed infill has been achieved through the indiscriminate and opportunistic subdivision of individual suburban lots by "mom and dad" investors (Cuff and Dahl 2009), which typically does not reduce car usage, erodes urban forests, and aggravates local communities (Bolleter 2016). Indeed, such approaches to infill development have led to community resistance (the NIMBY[3] factor), and have created what one council official has referred to as a "public sullenness" (Kelly and Donegan 2015).

1.1.5 Introducing Greenspace-Oriented Development

While the principles of TOD are well established and have some validity, it is our belief that we need a complementary strategy for achieving infill development. In this book, we propose Greenspace-Oriented Development (GOD). While TOD correlates urban densification with public transport hubs, GOD correlates urban densification with significant, upgraded public green spaces or parks that are relatively well served by public transport (Fig. 1.4). At its foundation, a GOD approach builds upon the now well-recognized importance of urban green spaces in delivering an array of benefits to urban dwellers, and most importantly, in underpinning approaches for greater sustainability and livability in cities.

[3]NIMBY "not in my back yard".

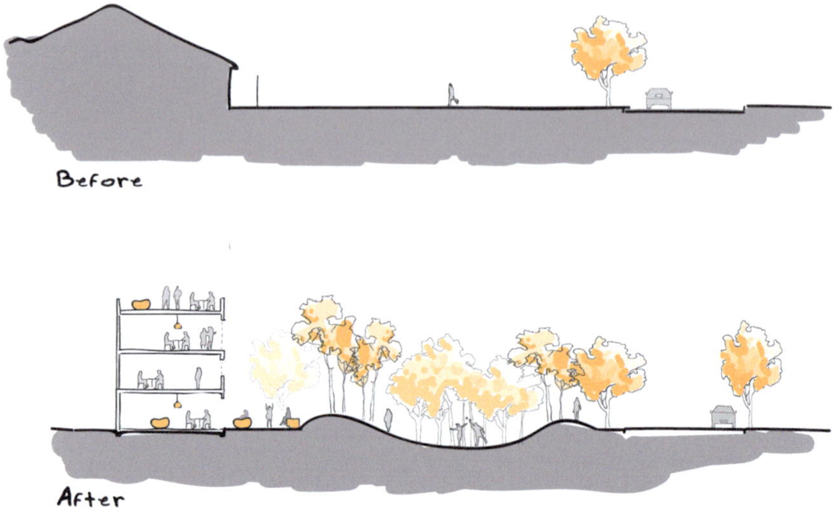

Fig. 1.4 Introducing Greenspace-Oriented Development (GOD): In this book, we propose GOD, a strategy that correlates urban densification with significant, upgraded public green spaces

In this book, we lay the foundations for this complementary strategy for infill development. We examine the potential advantages of enhancing the amenity and ecosystem services provided by parks in middle-ring, greyfield suburbs. Further, we examine how the employment of a social–ecological approach in the redesign of urban parks should uplift the real estate value of adjacent areas. In conjunction with rezoning, this should enable greater densification that is commercially viable for developers. We hypothesize that the redesign of urban parks towards greater naturalness,[4] ecological function, and diversity of active and passive recreational uses, offers an incentive for residents to support well-designed infill development at higher densities. We believe policymakers have overlooked the potential of such an approach as a key strategy for urban infill and revitalization.

1.1.6 How Does GOD Differ from Related Concepts?

The idea of using nature to improve the lives of those in cities is not new. More than a century ago, Frederick Law Olmsted used the natural processes of wetlands to guide his design of Boston's Green Necklace series of parks. In the early twentieth-century, urban thinkers conceived various utopian city models that reflected a clear concern about reconciling the city and nature, for example, the generous greenbelts,

[4]The term naturalness generally describes how close a landscape is to a perceived natural state, in a continuum between natural, intact landscapes (e.g. wilderness areas) and non-natural, highly modified landscapes (e.g. built-up areas) (Ode et al. 2009; Anderson 1991).

and the agrarian socialism of Ebenezer Howard's "Garden City" of 1902 (Weller 2017). In the 1960s, Ian McHarg mapped natural areas providing important services and protected them in his plans for urban development (McDonald 2015).

More recently, various urban design and planning theories have sought to reconcile urban form and green systems. *Green Urbanism* proposes a "city that maximizes landscapes, gardens and biodiversity" and green infrastructure[5] (Lehmann 2010). In a similar vein, *Landscape Urbanism* foregrounds landscape as the "ultimate system to which all goes, and from which all comes, a template for urbanism" (Weller 2006). Likewise, *Ecological Urbanism* proposes an apparently "new sensibility – one that has the capacity to incorporate and accommodate the inherent conflicts between ecology and urbanism" (Mostafavi 2010). Finally, *Biodiversity Sensitive Urban Design* is a protocol for urban design that aims to create a net benefit to native species and ecosystems through the provision of essential habitat and food resources (Garrad and Bekessy 2015).

GOD builds upon these worthy theories, exploring how good quality parks that are relatively well served by public transport and are surrounded by well-designed higher-density urban precincts, can help create and activate medium-density hubs in suburban areas. GOD parks not only provide a myriad of social and ecological benefits to residents, but they can also have a crucial role in activating the local neighbourhood, and instill a sense of, and attachment to, place.

Yet, GOD[6] (as we propose it) differs from other conceptual frameworks, such as Ecological Urbanism, in that it provides tools that bridge theory and practice. While theories such as New Urbanism provide various tools (for example, the Transect or From Based Codes) by which designers can implement theoretical concepts into practice; Landscape Urbanism, for instance, offers no such props for practitioners, leaving them to understand how theory should be implemented (Dennis and McIntosh 2013). Through this book, we aim to bridge this divide. This is important because globally, governance and the processes "by which things get built" are a major stumbling block to equitable and sustainable planning (Berger et al. 2017). GOD is also distinguished from other conceptual frameworks because the authors represent the otherwise often divergent fields of urban design and urban ecology—a gap we aim to narrow through this shared publication.

1.1.7 Who Is This Book Directed Towards?

This book is meant for a wide audience, from researchers working in urban planning, urban sustainability and livability, particularly in the context of urban densification,

---

[5]Green infrastructure refers to the ecological features and systems, from wetlands to urban forests, that provide a host of benefits to urban residents. These include storm water collection and treatment, climate moderation, and cleansing of air, among others (Newman et al. 2009).

[6]Readers should note the work of the Greater City Commission, and Rod Simpson in particular, to correlate urban density with greenspace. This high-level policy work has emerged in parallel with this book (Simpson 2018).

to practitioners involved in the planning, design and implementation of urban green spaces and residential areas. This includes planners, architects, landscape architects, urban designers, developers, policymakers, engineers, park managers, environmental and social science experts, amongst others.

1.2 Overview of the Book's Chapters

Earlier in this chapter, we set out the big issues facing suburban cities of the western world in relation to achieving urban densification. We have also introduced in brief the concept of GOD. Here we set out an overview of the book's subsequent chapters.

In Chap. 2, "Transit-Oriented Development and Its Problems", we set out the ascendant ideology of TOD and explore its relative failure to deliver infill outcomes in urban densification settings in Australian cities. We identify the key barriers to TOD, including land assembly and development feasibility issues, community resistance, a lack of consumer demand and infrastructure provision challenges. While these barriers are not new, emerging trends also threaten the delivery of TOD planning. For example, the predicted widespread adoption of Fully Automated Vehicles (FAVs) could mean that correlating urban densification with public transport will be of decreasing importance.

Given the challenges of achieving TOD, a substantial portion of urban densification is occurring in an ad hoc indiscriminate manner, a trend we explore in relation to Australian cities. Using Perth as a case study, we explore how comparatively modest land values and community hostility to higher density infill have favoured background infill. This chapter raises concerns about this indiscriminate form of urban densification, as it typically yields low amenity outdoor space, reduces urban forest cover, does not support the development of cultural amenity and conveniences, and provides minimal access to public transport.

An "infill good, sprawl bad" polarity, partly perpetuated by the planning community, pervades arguments about urban form in Australia. In contradistinction, we argue in this chapter that urban densification is delivering mixed results in respect to urban livability and that planners require viable alternatives.

In Chap. 3, "Why GOD? The Benefits of Greenspace-Oriented Development", we set out a complementary strategy for urban densification, Greenspace-Oriented Development (GOD). Through GOD, we propose to weave together the benefits of urban green spaces with the positive aspects of suburbia—access to open space and nature—and those of high-quality medium-density urban infill—public transport, community facilities and good urban design. As mentioned before, GOD correlates urban densification with significant, upgraded public green spaces or parks that are relatively well served by public transport. The upgraded parks, and their connecting streetscapes, operate as a multifunctional, communal "backyard" for residents living in a surrounding higher-density urban precinct.

In this chapter, we describe GOD, its key principles and the settings to which it is most suited. We explore the main benefits green spaces can offer in the context of increased density—to people, biodiversity and the local environment. We also explore the benefits that well-designed, higher-density urban precincts can offer to the parks themselves.

In Chap. 4, "A Manual for Implementing Greenspace-Oriented Development", we suggest a detailed process to guide GOD implementation. Overall, we explain how the upgrade of suitable urban parks—of a required size and accessibility to public transport systems—combined with the rezoning of the adjacent urban precinct can lead to increasing land values and catalyze redevelopment. Specific steps include (1) select parks for upgrading; (2) upgrade parks; (3) rezone the urban precincts surrounding parks; (4) catalyze and facilitate redevelopment; (5) decentralize services infrastructure; (6) conduct needs-based assessment and equip parks; and (7) upgrade surrounding key streetscapes. Through this step-by-step process, we set out how GOD can guide increased urban densification, along with the redesign of public green spaces to offer multiple benefits to park users and local residents.

Accompanying each of the steps are diagrams that will aid planners and policymakers in developing a policy for urban densification. Moreover, we have produced three-dimensional visualizations that people can imagine themselves into and planners can use to kick-start community engagement exercises. As Kim Dovey explains, one of the major blockages to "transformational change" in Australian cities has been a "lack of design vision that captures the public imagination for more sustainable urban futures" (Dovey and Woodcock 2014). This book aims to tackle this lacuna head-on.

In Chap. 5, "Conclusion", we briefly summarize the main findings of the book, consider key implications, and direct researchers to potentially fertile areas for future research.

1.3 Conclusion

Robert McDonald, Lead Scientist at The Nature Conservancy, has stated that the most successful cities in the twenty-first century will do the best job of protecting the essential ecosystem services nature provides to humans (McDonald 2015). While Australian state planning agencies have sought through urban densification, to protect the ecosystem services provided by peri-urban green systems, in many instances they are failing. This is primarily because a love for suburban living runs deep in the psyche of Australian people. In response to this situation, this book sets out a complementary approach for densifying Australian cities—one that works with, rather than overrides, Australia's prevailing suburban sensibility.

References

Anderson JE (1991) A conceptual framework for evaluating and quantifying naturalness. Conserv Biol 5(3):347–352. https://doi.org/10.1111/j.1523-1739.1991.tb00148.x
Bento AM, Cropper ML, Mobarak AM, Vinha K (2005) The effects of urban spatial structure on travel demand in the United States. Rev Econ Stat 87(3):466–478. https://doi.org/10.1162/0034653054638292
Berger A, Kotkin J, Guzman C (2017) Introduction. In: Berger A, Kotkin J, Guzman C (eds) Infinite suburbia. MIT, Boston, pp 10–23
Bolleter J (2016) Background noise: a review of the effects of background infill on urban liveability in Perth. Aust Plan 10:1–14
Bolleter J (2017) Fringe benefits? A review of outer suburban development on Perth's fringes in relation to state government goals concerning the natural environment and efficient transport connectivity. Aust Plan 54(2)
Bolleter J, Weller R (2013) Made in Australia: the future of Australian cities. University of Western Australia Publishing, Perth
Brownstone D, Golob TF (2009) The impact of residential density on vehicle usage and energy consumption. J Urban Econ 65(1):91–98. https://doi.org/10.1016/j.jue.2008.09.002
Bruegmann R (2017) The anti-suburban crusade. In: Berger A, Kotkin J, Guzman C (eds) Infinite suburbia. MIT, Boston, pp 26–37
Calthorpe P, Fulton W (2001) The regional city. Island Press, Washington
City of Melbourne (2010) Transforming Australian cities for a more financially viable and sustainable future. City of Melbourne, Melbourne
Cuff D, Dahl P-J (2009) Rx for the R1: sustaining the neighbourhood. In: Bolchover J, Solomon J (eds) Sustain and develop. 306090, Inc, New York, pp 24–33
Curtin University, Hames Sharley (2013) The housing we'd choose: a study for Perth and Peel. Department of Housing, Department of Planning, Perth
Dennis M, McIntosh A (2013) Landscape and the city. Landscape urbanism and its discontents: dissimulating the sustainable city. New Society Publishers, Gabriola Island
Dittmar H, Belzer D, Autler G (2004) An introduction to transit-oriented development. In: Dittmar H, Ohland G (eds) New Transit Town: best practices in transit-oriented development. Island Press, Washington
Dodson J, Sipe N (2008) Unsettling suburbia: the new landscape of oil and mortgage vulnerability in Australian cities. Griffith University, Brisbane
Dovey K, Woodcock I (2014) Intensifying Melbourne: transit-oriented urban design for resilient urban futures. Melbourne School of Design, The University of Melbourne, Melbourne
Forster C (2006) The challenge of change: Australian cities and urban planning in the new millennium. Geogr Res 44(2):173–182
Garrad G, Bekessy S (2015) Biodiversity sensitive urban design: creating urban environments that are good for people and good for nature. RMIT. https://ggarrardresearch.wordpress.com/biodiversity-sensitive-urban-design/. Accessed 16 June 2019
Hagan S (2017) Metabolic suburbs or the virtue of low densities. In: Berger A, Kotkin J, Guzman C (eds) Infinite suburbia. MIT, Boston, pp 468–477
Kelly J-F, Donegan P (2015) City limits: why Australian cities are broken and how we can fix them. Melbourne University Press, Melbourne
Kelly J-F, Breadon P, Reichl J (2011a) Getting the housing we want. Grattan Institute, Melbourne
Kelly J-F, Weldmann B, Walsh M (2011b) The housing we'd choose. Grattan Institute, Melbourne
Kelly J-F, Breadon P, Davis C, Hunter A, Mares P, Mullerworth D, Weidmann B (2012) Social cities. Grattan Institute, Melbourne
Lehmann S (2010) The principles of green urbanism: transforming the city for sustainability. Earthscan, UK
McDonald R (2015) Conservation for cities: how to plan and build natural infrastructure. Island Press, Washington

McNeill JR, Engelke P (2016) The great acceleration: an environmental history of the anthropocene since 1945. Harvard University Press, Cambridge

Mostafavi M (2010) Why ecological urbanism? Why now? In: Mostafavi M, Doherty G (eds) Ecological urbanism. Lars Muller Publishers, Baden

Nechyba TJ, Walsh RP (2004) Urban sprawl. J Econ Perspect 18(4):177–200. https://doi.org/10.1257/0895330042632681

Newman P, Beatley T, Boyer H (2009) Resilient cities. Island Press, Washington

Ode Å, Fry G, Tveit MS, Messager P, Miller D (2009) Indicators of perceived naturalness as drivers of landscape preference. J Environ Manag 90(1):375–383. https://doi.org/10.1016/j.jenvman.2007.10.013

Radeloff VC, Stewart SI, Hawbaker TJ, Gimmi U, Pidgeon AM, Flather CH, Hammer RB, Helmers DP (2010) Housing growth in and near United States protected areas limits their conservation value. Proc Natl Acad Sci 107(2):940–945. https://doi.org/10.1073/pnas.0911131107

Schneider A, Woodcock CE (2008) Compact, dispersed, fragmented, extensive? A comparison of urban growth in twenty-five global cities using remotely sensed data, pattern metrics and census information. Urban Stud 45(3):659–692. https://doi.org/10.1177/0042098007087340

Seto KC, Kaufmann RK, Woodcock CE (2000) Landsat reveals China's farmland reserves, but they're vanishing fast. Nature 406(6792):121–121

Seto KC, Guneralp B, Hutyra L (2012) Global forecasts of urban expansion to 2030 and direct impacts on biodiversity and carbon pools. PNAS 109(40):16083–16088

Simpson R (2018) Start with landscape. Australian Institute of Landscape Architects. http://www.aila.org.au/iMIS_Prod/AILAWeb/Media_Releases/AILA_Calls_for_a_National_Green_Infrastructure_Strategy.aspx. Accessed 28 June 2019

Sturm R, Cohen DA (2004) Suburban sprawl and physical and mental health. Public Health 118(7):488–496. https://doi.org/10.1016/j.puhe.2004.02.007

United Nations (2017) The sustainable development goals report 2017. United Nations, New York

United Nations (2018) World urbanization prospects: the 2018 revision [key facts]. Department of Economic and Social Affairs, Population Division

United Nations General Assembly Draft Outcome Document of the United Nations Conference on Housing and Sustainable Urban Development (Habitat III), 2016. Document A/CONF

Wall A (2017) Sprawl is dead: long live the low-density city. In: Berger A, Kotkin J, Guzman C (eds) Infinite suburbia. MIT, Boston, pp 571–594

Weller R (2006) Global theory, local practice. Kerb 15:66–71

Weller R (2017) The city is not an egg: western urbanization in relation to changing conceptions of nature. In: Steiner F, Thompson G, Carbonell A (eds) Nature and cities: the ecological imperative in urban design and planning. Lincoln Institute of Land Policy, Cambridge, pp 31–49

Woodcock I, Dovey K, Wollan S, Beyerle A (2010) Modelling the compact city; capacities and visions for Melbourne. Aust Plan 47(2):94–104

Zhao Z, Kaestner R (2010) Effects of urban sprawl on obesity. J Health Econ 29(6):779–787. https://doi.org/10.1016/j.jhealeco.2010.07.006

Chapter 2
Transit-Oriented Development (TOD) and Its Problems

Abstract In this chapter, we set out the ascendant ideology of TOD and review the barriers to its implementation in urban densification settings in suburban cities. These barriers can include land assembly and development feasibility issues, community resistance, lack of consumer demand and infrastructure provision challenges. As a result of these barriers, attempts to implement TOD have proven a challenge. Moreover, a substantial portion of urban densification is occurring in an ad hoc manner, a trend we explore in relation to Australian cities. This chapter raises concerns about this indiscriminate form of urban densification, as it offers minimal access to public transport, contributes little to urban activation, compromises urban forests, and entrenches community opposition to infill development. Given these issues, this chapter stresses the need for a complementary strategy to support urban densification in suburban cities.

Keywords Transit-Oriented Development · Urban consolidation · Infill development · Greyfields · Public transport · Sustainability · Urban livability · Public open space · Urban parks

2.1 An Overview of TOD

Transit-Oriented Development (TOD) planning aims to concentrate urban development—including high-density residential and office/commercial development—in precincts around public transport hubs (usually within an 800 m, 10-minute walk), in order to increase public transport use and deliver urban infill (Curtis 2012). Because of its apparent logic and relative simplicity (at least in concept), planners are pursuing TOD strategies in many cities throughout the world (Curtis et al. 2010). Given the ascendency of TOD ideology, this chapter provides a timely critique regarding its effectiveness in delivering urban densification in Australian cities. Nonetheless, readers should be mindful that achieving urban densification in suburban cities is a difficult thing. We do not mean to denigrate the sincere efforts of practitioners to achieve TOD, instead to point out the need for complementary strategies.

J. Bolleter and C. E. Ramalho, *Greenspace-Oriented Development*, SpringerBriefs in Geography, https://doi.org/10.1007/978-3-030-29601-8_2

The idea that transit might orient development is certainly not new. In this respect, Peter Calthorpe considers himself "a reviver rather than an originator of ideas" (Carlton 2009). For example, Ebenezer Howard's satellite cities were to be enabled by rail transit access. As he described in the early twentieth century, his planning concept was focused on rail networks as the primary conduit between areas of urban development (Carlton 2009). Nonetheless, Peter Calthorpe codified the idea of Transit-Oriented Development (TOD) in the late 1980s, and TOD became a central tenet of modern planning when Calthorpe published "The New American Metropolis" in 1993 (Carlton 2009).

Globally, TOD proponents claim that compact urban form co-located with public transport nodes will deliver a multitude of benefits. These benefits include making public transportation more economically viable (Hagan 2017), increased mobility of low-income households, a boost to local services, and less automobile dependency, which in turn can help reduce energy consumption and greenhouse gas emissions, and encourage walking (Congress of New Urbanism 2016). Proponents believe that TOD will also increase the efficiency of infrastructure investment, and reduce the cost of maintenance, particularly for line systems, such as energy and water supply, and waste disposal. Finally, advocates for TOD believe it will provide residents with a diversity of local jobs, stimulate knowledge diffusion and thus economic growth, all of which contribute to a higher quality of life for residents (OECD 2012). Because of the horde of perceived benefits, planners have widely accepted the ideology of TOD. As a result, "now almost every metropolitan region with major public transport infrastructure has adopted some form of high-density TOD scenario" (Carlton 2009).

2.2 TOD Planning in Australian Cities

Reflecting the ascendency of TOD ideology, all Australian State and Territory capital cities plan to achieve urban densification around public transport nodes (Department of Infrastructure 2017; Department of Planning 2017; Department of Planning and Western Australian Planning Commission 2015; Victoria State Government 2017). Through such development, these plans attempt to avoid Australian cities sprawling in what is recognized as a typically unhealthy, socio-economically stratified, unsustainable and unproductive manner (Kelly and Donegan 2015). To this end, across the nation, city planning policies, on average, stipulate that 60% of all new residential development should be infill, yet less than that is typically being achieved (Bolleter and Weller 2013).

The flagships of state government planning policy for urban densification are Activity Centres. Undergirding the planning of these Activity Centres are the principles of TOD. Activity Centres policies are defined here as policies that seek to cluster civic, commercial and mid- to high-density residential land uses around public transport nodes, rather than allowing them to disperse in an unplanned manner wherever the market chooses to locate them (Goodman and Moloney 2004). In a national effort to transition from monocentric to polycentric urban systems, State and Territory policies in Australia have identified 343 Activity Centres for infill development nationwide (Bolleter and Weller 2013).

2.3 What Are the Predominant Barriers to TOD in Australian Cities?

When considered at the metropolitan scale, the co-location of residential density, commercial activity and public transport found in Australia's Activity Centre planning appears common sense, but on the ground, it confronts many barriers to implementation. As a result, relative to other real estate investments, TODs are often more complex, take more time, are riskier and are ultimately more expensive (Carlton 2009). To explain this situation, we will discuss these barriers as they relate to TOD in urban densification settings, in four broad categories: community barriers, development feasibility barriers, governance barriers and emerging barriers.

2.3.1 Community Barriers

Community-related barriers to TOD in urban densification settings are twofold. First, existing residents are often unreceptive to increases in urban density, and second prospective residents often do not aspire to living in Activity Centres.

Development within existing Activity Centre sites has sometimes been difficult because of community resistance. Density increases are often perceived as a threat to the suburban life (Dovey and Woodcock 2014). As Wendy Sarkissian tells us, "A huge battle has been waging for more than two decades about this matter in Australia…" (Sarkissian 2013)—and that a public "sullenness" exists in relation to urban infill in suburban neighbourhoods (Kelly and Donegan 2015). As Kim Dovey explains, our communities are saying to us, "loudly and forcefully, we don't want 4/5/6 storeys in our Activity Centres, we would like to retain the two storey streetscape…" (Dovey and Woodcock 2014). Communities often convey several key reasons for this opposition. These include fears related to perceived increased traffic and parking problems (Parliamentarian 2018; Local Government planner 2018), declining property prices (Development Institute representative 2018), and a perception that trains are already overcrowded and that TOD will just make a bad situation worse (Rice 2016). These are compounded by concerns about a lack of privacy and amenity (Parliamentarian 2018), the destruction of urban forests (Community Representative 2018), loss of heritage and neighbourhood character, the feeling that TOD developments don't belong in their neighbourhood, and finally cynicism that TOD is just about developers "making a lot of money at their community's expense" (Rice 2016).

While there is ample evidence of community resistance to Activity Centre development, there is comparatively little evidence that prospective residents aspire to living in Activity Centres (Holling and Haslam McKenzie 2010). A study based in Perth entitled "The Housing We'd Choose" provides some insights, whereby researchers explored the relative importance of a wide range of housing attributes so as to establish what households desired (Curtin University and Hames Sharley 2013).[1] Significantly, 70% of survey respondents ranked being "near a shopping centre" as the most important dwelling attribute. This reveals that dwellings in Activity Centres planned around shopping centres are potentially desirable, at least in this respect. However, over 40% of respondents believed housing located away from a railway line was important (Curtin University and Hames Sharley 2013), probably due to concerns about train noise, privacy and commuter car parking (Holling and Haslam McKenzie 2010). Moreover, 65% of respondents regarded a dwelling being "away from busy roads" as being important (2013), this dwelling attribute ranking as the fourth most important. Major roads bisect many of Perth's proposed Activity Centres. The benefits of an urban lifestyle as promoted in Activity Centres also do not seem to be that popular. For instance, being near cafes and restaurants ranked 12th, easy access to the city (through public transport) ranked 22nd, having a range of local employment opportunities also ranked 35th and easy access to bars/pubs and nightlife ranked 39th. In Perth, at least, people do not seem to desire the benefits of TOD as much as policy makers have anticipated.

2.3.2 Development Feasibility Barriers

The development of TOD-driven Activity Centres faces many other barriers in addition to a relative lack of community buy-in. Train station surroundings often have heritage building stock, are typically complex "knots" of intersecting road and rail infrastructure and have fragmented land ownership. The complex array of vested interests attached to these factors is a major constraint in Activity Centre planning (Murphy 2012). Compounding this is that there remains a high demand for expansive car parks adjacent to public transport so that people can "park and ride", which conflicts with attempts to create walkable and activated Activity Centres (Holling and Haslam McKenzie 2010). Moreover, Activity Centre sites often do not have adequate service infrastructure and the costs of upgrades can affect the feasibility of a development (Rowley and Phibbs 2012). The related uncertainty around developer contributions and "who will pay" is a major impediment to TOD infill development (Rowley and Phibbs 2012).

[1] A subset of this study was an online survey entitled "What Matters Most" in which researchers asked 866 people to rate the features of a home they placed the highest priority on. Respondents ranked 76 attributes arranged into five broad categories: convenience, local amenities, local environment, dwelling design and dwelling features.

2.3.3 Governance Barriers

TOD-driven Activity Centre development requires the participation of many actors and occurs in a fragmented regulatory environment, adding complexity, time, uncertainty, risk and cost to projects (Dittmar et al. 2004). Challenges to Activity Centre development arise from the political structure that realizes infill development. The state government sets infill targets for local government areas, yet local governments carry much of the responsibility for infill development decisions (Dovey and Woodcock 2014). Moreover, local governments are frequently "elected to enforce the anti-development views of their residents" and, as such, disperse as much infill development in a form that is as palatable to existing residents as possible (i.e. in a low- to medium-density and dispersed pattern) (Dovey and Woodcock 2014). Finally, local councils are typically not staffed to adequately navigate the complications of TOD (Dovey and Woodcock 2014) or funded to deliver the level of investment Activity Centre's require (Gray et al. 2010).

Moreover, while there has been substantial investment in passenger rail since the 1990s, the dominance of transport planning over land-use planning—at the state government level—is evident in planning for rail networks. The primary focus has generally been on transport function, and only comparatively recently has there been a real attempt to integrate the transport network carefully with land-use activity (Curtis 2010). The result is often a lack of land-use transport integration—isolated public transport hubs that are beyond walking distance from low-density residential areas (Curtis 2010).

2.3.4 Emerging Barriers

While TOD-driven Activity Centre implementation already faces considerable barriers, other potential challenges are emerging. For example, with the arrival of self-driving cars it is possible that the line separating public and private transportation will erode as users summon and share vehicles of varied sizes for different kinds of trips (Bruegmann 2017). It is quite possible that this kind of shared vehicle, allowing direct movement from any point A to point B, will lessen the requirement for public transportation in the form of buses and trains, and will work against the delivery of compact cities (Falconer et al. 2016). As with many other advances in technology over the last century, these developments could allow people more freedom to choose exactly what kind of environment they would prefer to live in (Bruegmann 2017). This, in all likelihood, would be low-density suburban settings (Hagan 2017).

Another emerging transport type, "trackless trams"—referred to as Autonomous Rail Transit (ART)—also potentially reduces the need for TOD-driven Activity Centre development. ART vehicles are based on technology developed in Europe and China by taking technology from high-speed rail and utilizing it in a bus (Newman 2018). The result is essentially an electric bus that has the speed, capacity and ride qualities of light rail with its "autonomous optical guidance system, train-like bogies with double axles and special hydraulics and tyres" (Newman 2018). These ART vehicles can also move significant numbers of people. The typical three-carriage system can accommodate 300 people (Newman 2018). Finally, ART vehicles can also sidestep the worst aspects of light rail/trams, namely, disruption and cost. It can take years to lay tracks for light rail/trams, causing major upheaval to local businesses (Newman 2018).[2] The result of such a transformative transport type could be ART vehicles travelling down major roads, effectively servicing broader suburban areas with efficiency and flexibility. Henceforth, knots of urban density around heavy rail stations—for instance—may be superfluous.

2.4 What Is Our Record for Delivering TOD in Australian Cities?

While contemporary TOD principles originated in the US, attempting to implement TOD there has proven a challenge. Not surprisingly given the barriers, only a small percentage of the anticipated TOD projects have been developed, and what has been built does not fully reflect Calthorpe's original vision (Carlton 2009). As Hank Dittmar explains in relation to an overarching North American study, "Sadly, our review of the projects that are emerging across the country reveals that many of the first phases of these new 'transit towns' fail to meet [TOD's] objectives" (Dittmar et al. 2004). As Ian Carlton declares, little has actually changed in the behaviour of Americans or their preferences for transport to signify a desire for TOD in comparison to car-oriented suburbia (Carlton 2009). As a result "few, if any, TODs have been developed without public subsidies" (Carlton 2009). In today's restricted funding environment, this means that many TOD developments will not be feasible.

The situation is similar in Australia—with some isolated exceptions. As a result of the barriers to TOD we set out earlier, there exists a "divergence between the compact city imagined in metropolitan plans and what is occurring on the ground in Australian cities" (Gray et al. 2010). The reality of urban development contrasts starkly with the TOD vision. Clive Forster makes the point that this simplistic vision of metropolitan sustainability is contradicted by the structures of our cities, which remain differentiated and dispersed rather than neatly multi-nucleated (Gleeson et al. 2010). Despite the application of TOD thinking in Perth, Adelaide, Melbourne, Sydney and South East Queensland, full implementation has proven a challenge (Kelly and Donegan 2015; Goodman and Moloney 2004; Burton 2017; Goodman 2017; Randolph et al. 2017). This is despite the fact that such policies have been in place for a considerable amount of time, in some cases since the 1980s (Murphy 2012).

[2] Indeed, Sydney's light rail project has costed over $120 million per km. In contrast, the infrastructure for ART vehicles apparently costs only $6–8 million per km, and proponents optimistically believe "it can be put into a road system over a weekend" (Newman 2018).

As Jago Dodson explains, "…despite more than two decades of densification policy, across Australia's major cities there are vast suburban regions of low density development" (2010). Indeed, Australian cities have some of the lowest population densities in the world—Melbourne, Adelaide, Perth and Brisbane averaging only 16, 14, 12 and 9 people per ha, respectively (Hurley et al. 2017). Moreover, the 2016 Census of Population and Housing found that only 10% of all people in Australia spent Census night in an apartment (Australian Bureau of Statistics 2017). The evidence shows that despite the rhetoric and agenda of planning authorities and experts, the suburban home remains the preferred choice of families (Elliot 2017).

Detractors of suburban living point to rapidly rising apartment projects in inner city areas in their quest for evidence to prove that Australians are now desiring high-density housing over the suburban alternative (Elliot 2017). Indeed, Elliot explains that the "rapid escalation in apartment construction in Australia's capital cities is little more than a frenzy of speculative investment" (Elliot 2017). The data partly supports such claims. The 2016 Census showed that for all apartments across Australia, well over half (59%) of tenants were renting. In contrast, in the same survey, only 21% of separate houses were rented (Australian Bureau of Statistics 2017).

Private vehicle ownership and use provides further evidence for the relative failure of TOD planning. In 2016, nearly half (47%) of households living in apartments had one registered motor vehicle—generally parked on-site, and 16% of apartment households upheld a "two-car" lifestyle (Australian Bureau of Statistics 2017). While the total distance driven and total number of car trips people are taking in Australian cities are growing more slowly than in previous decades, car use is still growing overall (Kelly and Donegan 2015). Flexibility and convenience, combined with the car-friendly structure of our major cities, means private vehicles account for almost 90% of passenger kilometers travelled (Infrastructure Australia 2018). Even with substantial mode shift driven by new investment or policy changes, this is unlikely to change significantly in the coming years (Infrastructure Australia 2018).

Problems in delivering urban densification in relation to public transport hubs have (in part) led to respected commentators, such as the former commissioner of the National Capital Development Commission, Tony Powell, to describe Australia's metropolitan planning as "a sad parade of failing capital city strategic plans" more concerned with public relations than planning (Elliot 2017). To further illustrate this point, in the following section, we will set out a brief history of attempts to deliver TOD in Australia's major capital cities.

2.4.1 Sydney

Sydney has been planning for urban densification since the 1980s (Randolph et al. 2017). In conjunction with high land costs and substantial population growth, this has seen the mix of dwellings across Sydney change significantly in the last two decades. There has been a decline in detached homes and increasing amounts of medium- and high-density apartment buildings, well above national averages (Randolph et al.

2017). Nevertheless, Sydney is a low-density city by international standards, and suburban housing remains the city's dominant built form, with 57% of the population living in detached homes (Infrastructure Australia 2018).

Moreover, there have been substantial challenges in delivering urban densification associated with public transport hubs. An assessment by planning consultant Pat Fensham in 2015 is less than flattering. He gave recorded "fails" on several key issues, proactive public transport initiatives and unconvincing commitment to the polycentric city agenda (i.e. Activity Centre networks) (Randolph et al. 2017). Further evidence is a New South Wales government report that acknowledged that from 2000 to 2010, developers had built only four in ten new homes in Sydney in "transit nodes" within 800 m of a train station or 400 m of a major bus stop or light rail station (Kelly and Donegan 2015). This pattern is certainly nothing new; the 1996 Census revealed that Sydney had the highest concentration of medium- and high-density housing unserved by rail services (Troy 2004). The New South Wales government report forecast that this pattern would continue to 2020 (Kelly and Donegan 2015).

2.4.2 Melbourne

Melbourne has been subject to many policies aimed at directing investment and development into designated Activity Centres. During the early 1980s, the metropolitan planning authority introduced a District Centre policy. By the mid-1980s, this policy was widely judged to have failed (Logan and McLoughlin in Goodman and Moloney 2004) and the Kennett state government gradually weakened and then abandoned it (Goodman 2017). A review in 1991 stated that support for the District Centre Policy had been ad hoc and called for a coordinated and renewed effort to make the policy work (Goodman and Moloney 2004).

Despite such failures, planners continued to base subsequent policy documents on TOD ideology. A key element of the Melbourne 2030 plan, released in 2002, was the designation of Activity Centres as appropriate foci for commercial, retail and office development, and higher density housing. Melbourne 2030 identified a hierarchy of five categories of centres, from the central city to neighbourhood centres (Goodman 2017). It named 114 centres comprising 25 principal, 79 major and 10 specialized centres, and proposed an incredible 900 neighbourhood centres throughout the metropolitan region (Goodman 2017).

Unsurprisingly given its lack of focus, Melbourne 2030 generated ample debate around its "failure to deliver on what it had promised" (Goodman 2017). Five years after its publication, an Expert Panel conducted an audit of progress (Audit Expert Group Goodman 2017). It revealed that the proportion of new greenfield development had continued to rise steeply and concluded that "on-the-ground" implementation was deficient in important areas, including the redirection of residential growth to existing areas, and increased development in Activity Centres (Goodman 2017). Indeed, the list of Activity Centres was far too extensive to provide a focus for investment in urban densification (Goodman and Moloney 2004; Goodman 2017).

As a result, some commentators described Melbourne 2030 as "superficial to the point of ridiculousness" (Elliot 2017).

An Australian Housing and Urban Research Institute (AHURI) study found that the overall amount of new housing built within 1 km of a major Activity Centre did not increase substantially after the introduction of Melbourne 2030 up until 2007 (Goodman et al. 2004, pp. 45–46). The study also found no increase in the proportion of new housing located within 1 km of train stations across the metropolitan area, another ambition of Melbourne 2030 (Goodman 2017). For example, in the City of Monash, in southeastern Melbourne, more than 98% of housing developments built from 2000 to 2006 were between two and seven dwellings. These smaller developments provided around nine out of ten new homes in that area (Kelly and Donegan 2015), yet most were not within walking distance of public transport.

As a result of these relative policy failures, Melbourne exhibits strong differentiation in housing supply between large detached suburban houses (with four or more bedrooms) in greenfield areas and new apartments, most which are small with one or two bedrooms, located in the central city and inner suburbs (Goodman 2017). Despite their numbers, a recent Melbourne study concluded that as many as one in five apartments were vacant, leading to the description of new projects as "ghost towers" (Elliot 2017). This is partly reflected in the data that indicates 70% of Melbournians live in detached homes and only around 15% live in apartments (Infrastructure Australia 2018). This may be because the latter do not suit many households, who would like to be able to choose semi-detached homes or low-rise apartments in established inner and middle suburbs, very few of which are being built (Kelly and Donegan 2015).

The current plan, Plan Melbourne 2017–2050, recently released by the Victorian state government (Victoria State Government 2017) provides new names and classifications for Activity Centres. It ignores the strident criticism of the previous policy that there were "too many named centres to be meaningful", and included all the principal and major centres from the superseded plans (Goodman 2017). In summary, while a wide range of stakeholders have spent substantial time and effort to produce grand plans for TOD in Melbourne, in the end, they have had much less effect than ambition (Goodman 2017; Kelly and Donegan 2015).

2.4.3 South East Queensland

South East Queensland denotes the conurbation between Brisbane, the Sunshine Coast and the Gold Coast. In response to the sprawling nature of this conurbation, planners in the region have been striving for urban densification for decades. Planners have based these strategic visions on TOD ideology and envisaged a region of "inter-connected communities that were more self-contained in terms of services and employment and as such, would generate less demand for travel in private vehicles" (Burton 2017).

Reflecting this, South East Queensland's 2005 plan proposed an infill target of 40% (South East Queensland Regional Organisation of Councils 2005), the 2009 plan a target of 50% (Department of Infrastructure and Planning 2009) and the current plan a target of 60% (Burton 2017; Department of Infrastructure 2017). While these are relatively modest in comparison with some state government infill targets, many in the development industry remain skeptical about achieving them, pointing to barriers to infill development such as the lack of substantial development sites, the challenges of site assembly and higher construction costs, all of which lead to higher prices for new developments and reduced project feasibility (Burton 2017).

Although academics have recognized these series of regional plans and strategies as laudable (Gleeson and Steele, Minnery and Low Choy in Burton 2017), the various plans have also been subject to criticism, mostly because of their failure to deliver their strategic visions on the ground (Burton 2017). As Mike Gillen explains, "contemporary patterns of economic development, housing and travel preferences do not correlate with a neat and ordered polycentric spatial form" (Gillen 2006). As with other Australian cities, the divergence between polycentric TOD-driven policy ambitions and the complex nature of urban transformations means the achievement of TOD "remains as elusive as ever for South East Queensland" (Burton 2017).

2.5 Feature Case Study: Perth

2.5.1 The Delivery of TOD in Perth

The principles espoused in Activity Centre planning have been (to some degree) a feature of Perth's planning since the mid-twentieth century. The Stephenson–Hepburn 1955 strategic plan aimed to create a series of "compact self-contained communities that had all the elements required for daily life" (Curtis 2010). The 1970s Corridor plan ventured an urban form comprising four corridors radiating from the central business district with regional centres at the ends, with the aim of reducing traffic congestion in the central city (Curtis 2010).

In 1988, the state government released Development Control Policy 1.6, "Residential Development Near Metropolitan Railway Stations", which aimed to promote higher density development close to train stations (Davis and Harford-Mills 2016). In a similar vein, in the 1990s "Metroplan", planners aimed to concentrate employment-generating activities and higher residential densities around public transport routes (Curtis 2010). Planners further extended these TOD principles in Perth's 2004 "Network City" plan. This plan comprised an extensive Activity Centre network and cast Perth as a "connected city" with higher densities around public transport nodes and areas of employment (Department of Planning and Western Australian Planning Commission 2015). Perth's "Directions 2031" (Department of Planning and Western Australian Planning Commission 2015) retained the TOD philosophy, as does the current plan "Perth and Peel @3.5 million" (Department of Planning and Western Australian Planning Commission 2015).

Despite Perth's most recent Activity Centre policy having been in place since 2010 (Government of Western Australia 2010), comparatively little infill development has occurred in the designated sites, except for Activity Centres such as Cockburn where state government owned much of the land (Figs. 2.1 and 2.2). Reflecting the challenges of delivering such development in Perth's Activity Centres, of the capital city regions, only Greater Brisbane had a lower population density than Perth. Indeed, detached houses accounted for 78% of Perth's total housing stock in 2011 (Maginn and Foley 2017).[3]

TOD has proven a challenge in Perth. As Curtis noted in 2012, policy translation from state government to local government has been "slow and inconsistent" and, despite considerable investment in public transport, it is evident that there has been little significant land-use change. As a result, Curtis identified an "implementation gap" between planning and on-the-ground development, and referred to the pace of change as "glacially slow" (Curtis 2012). The state government's failure to implement many public transport initiatives proposed within strategic plans for Perth (from 1955 to the present) has compounded this. Those include proposals for new rail, light rail, rapid bus and ferry systems (Davis and Harford-Mills 2016).

Perhaps as a result of these factors, Perth has had a 5-year decline in public transport use that has only recently been arrested (Acott 2019). Furthermore, "over the past 40 years, the number of cars in Perth has grown faster than its population, and it now has more cars per capita than any other Australian capital city, with some 83 vehicles per 100 people" (Department of Planning and Western Australian Planning Commission 2015). Reflecting this, in 2011, 77% of Perth residents drove to work (Department of Planning and Western Australian Planning Commission 2015).

The relative failure of TOD in Perth is despite the city has seen "one of the most deliberate attempts worldwide to move from car dependent development patterns to TOD" (Curtis 2012). Moreover, state planning policy has required TOD for decades, the public transport network has been progressively improved, institutional arrangements are strong (Curtis 2012), and Perth has been the beneficiary of successive mining-related boom periods.

2.5.2 *Background Infill*

A lack of development in Activity Centre sites in Perth has enabled substantial greenfield development, which in turn has destroyed vast swathes of remnant vegetation of the Southwest Australia global biodiversity hotspot,[4] exacerbated basic

[3]The infill rate for Perth was approximately 42% in 2017, up from 34% in 2015 (Department of Planning Lands and Heritage 2019); however, this is partly explained by patchy greenfield development in recent years.

[4]Between 2001 and 2009, suburban growth consumed an annual average of 851 ha of highly biodiverse land on the urban fringe (Weller 2009).

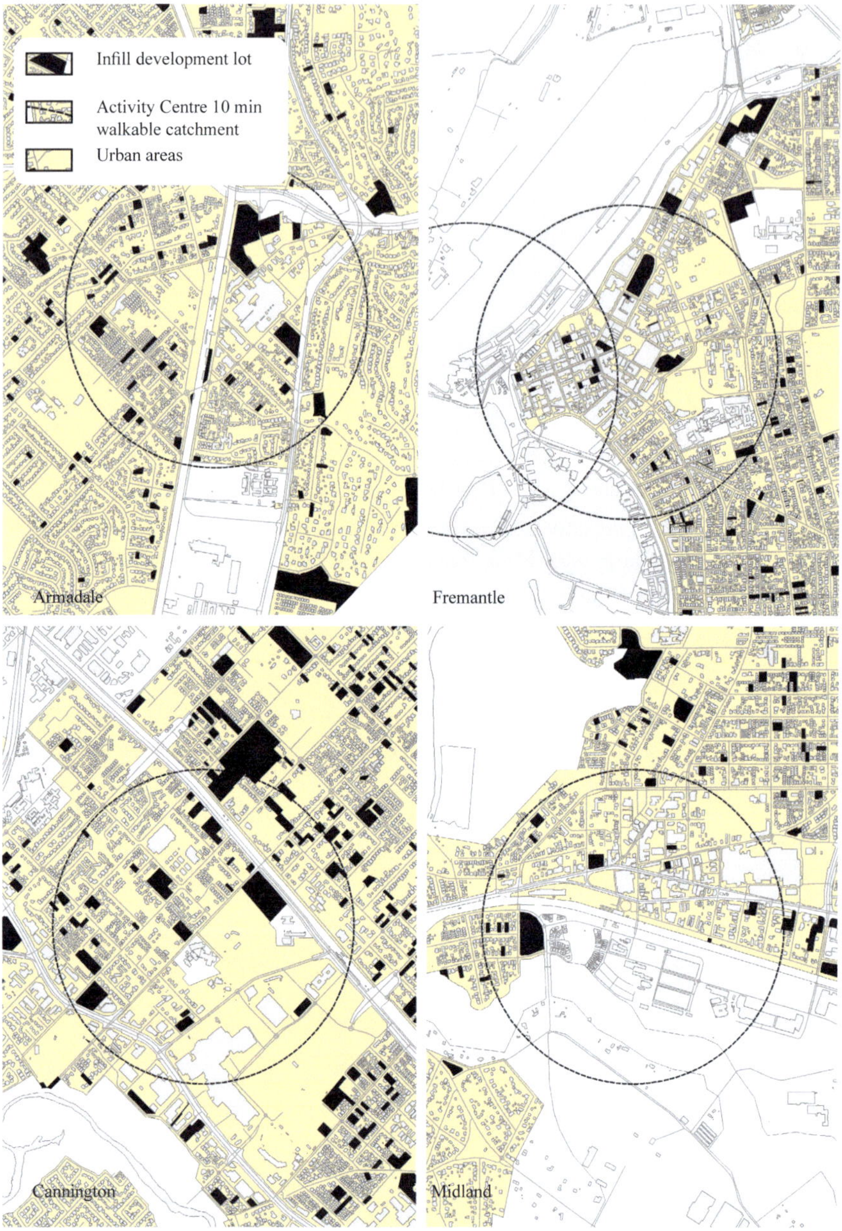

Fig. 2.1 The challenges of Activity Centre development: Mapping of Perth's Strategic Metropolitan Activity Centre's reveals comparatively little development between 2010 and 2019

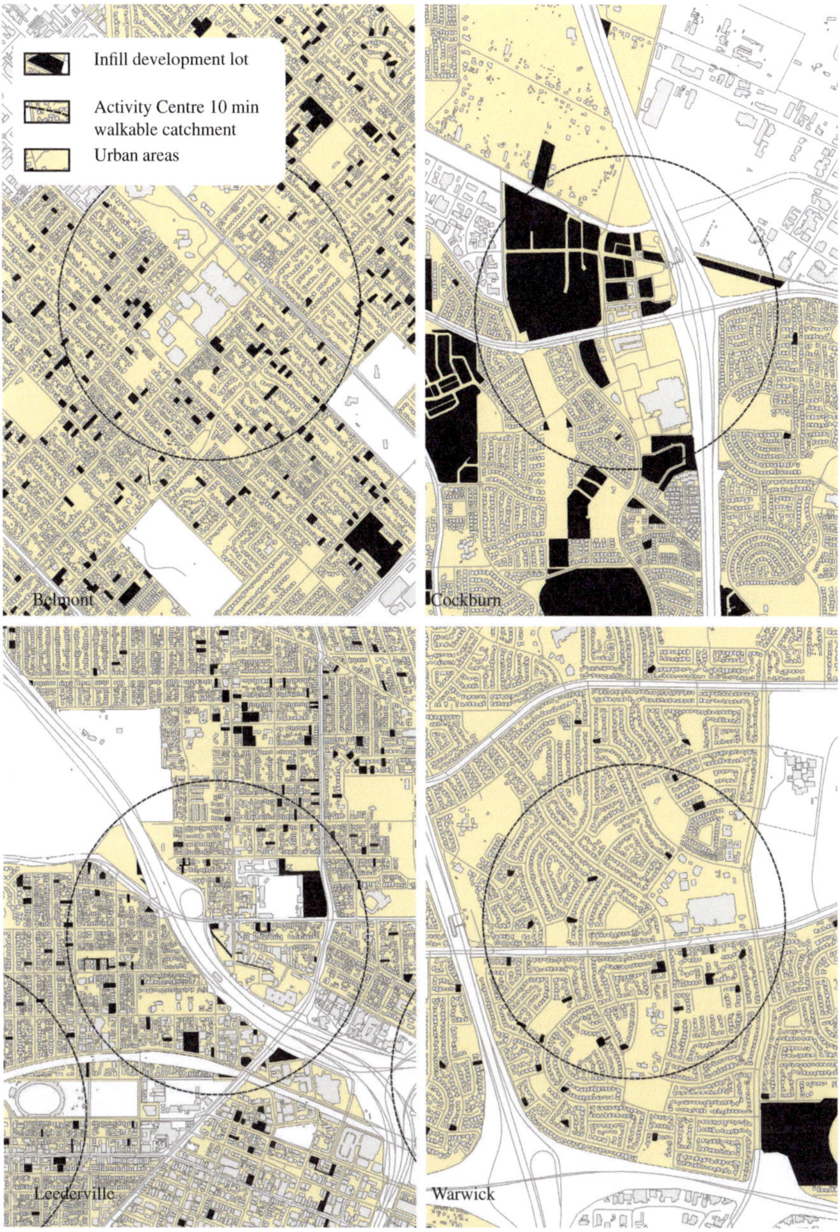

Fig. 2.2 The challenges of Activity Centre development: Mapping of Perth's Secondary Activity Centre's shows comparatively little development between 2010 and 2019, other than Cockburn which had substantial reserves of government-owned land

raw material shortages, and compromised Perth's food security by paving over agriculturally productive peri-urban land (Bolleter 2015). The challenges of TOD have also meant that a large proportion of infill development is occurring through the "do it yourself" subdivision of backyards. Planners refer to this form of infill development as "background" infill—namely, small projects yielding fewer than five group dwellings (Department of Planning Lands and Heritage 2019). Background infill is characterized by semi-detached, survey strata, group dwellings (generally single storey) organized around a communal driveway space leading to private garages adjacent to the dwellings (Fig. 2.3).

Perth's Urban Growth Monitor confirms the prevalence of background infill development. In 2017, infill projects where development resulted in between one to five new dwellings per lot (i.e. background infill) accounted for 56% of all the infill dwellings built in that year (Department of Planning Lands and Heritage 2019). The data for 2017 is, however, an improvement within the period 2012–2017, where background infill constituted 65% of all infill development. Nonetheless, despite decades of TOD planning in Perth, developers are delivering more infill in an ad hoc, "background" manner than in Activity Centres. This suggests there is value in reflecting on the type of infill development that developers are delivering (i.e. background infill), rather than what planners are aspiring to, but may not eventuate.

Fig. 2.3 Background infill: Background infill is characterized by up to five semi-detached dwellings around a communal driveway space leading to private garages. *Source* Nearmap

2.5.3 Impacts of Background Infill in Perth

In this section, we evaluate the impacts of background infill through a geospatial analysis. We have employed this approach to find correlations between sites of background infill and access to nature, retail destinations and public transport—all of which are crucial to urban livability (The Economist Intelligence Unit Limited 2012).

2.5.4 Access to Nature

"Contact with nature" is regarded as a universal requirement of "urban landscapes" (Arvola and Pennanen 2014), a sentiment also strongly shared by Perth's residents (Curtin University and Hames Sharley 2013). Not surprisingly, background infill development in Perth tends to provide much less garden area per person than an unsubdivided suburban form.[5] Compounding this situation, much of the open space that is provided in background infill is residual space, generated by the state government controlled Residential Design Codes (R-codes), which dictate a minimum 1.5 m setback between lot lines and building edges (State of Western Australia 2010). When "built out", this tends to result in narrow corridors of private open space (Fig. 2.4). A proclivity for private car parking adjoining private dwellings also means that, in many cases, 40% of the lot area relates to car parking and movement.

Planning has only partly compensated for this loss in the public realm. Most background infill is not located nearby regional open space, which provides the most authentic experience of nature in the city (Fig. 2.5). While middle-ring local government areas provide a substantial 40 m^2 of public open space per person (Fig. 2.6), it tends to be of poor quality and is often not well utilized (Bolleter 2015)—evidence of which readers can find in data measuring physical inactivity levels in residents (PHIDU Torrens University Australia) (Fig. 2.7). This is, at least partly, because across Perth's inner and middle-ring suburbs, where most background infill is occurring, parks are typically turf expanses sometimes having only scattered mature trees.[6]

Despite the significant amount of background infill development that has occurred to date, there has been no systematic upgrade of the public domain to service residents living at higher densities. The failure of local governments to upgrade open spaces is worrying because research tells us that higher-density residents have different needs from public space than their lower-density counterparts (Syme et al. 2001).

[5]The "classic" quarter acre block provided approximately 1,000 m^2 per household (Seddon 1994).

[6]Reflecting this, among the parks in Perth's inner and middle suburbs, 22% have no trees, only 10% have significant wildlife function and only 1% have wetlands (despite the fact that Perth was historically a landscape of wetlands). Furthermore, 74% of parks have a pervasive underlay of reticulated turf poorly suited to Perth's drying climate, 54% have no walking paths, only 16% have picnic tables and 9% have barbecue facilities (Centre for the Built Environment and Health 2013).

Fig. 2.4 Background infill: Background infill often results in narrow corridors of unusable private open space. *Source* Julian Bolleter

Fig. 2.5 Background infill and regional open space: Most background infill is not located near regional open space, which provides the most authentic experience of nature in the city

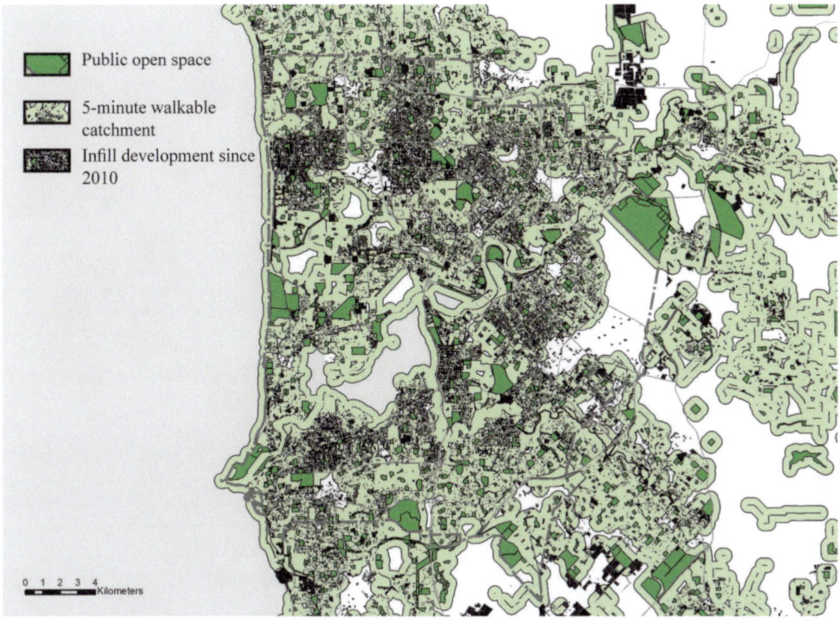

Fig. 2.6 Background infill and public open space: While middle-ring local government areas provide a substantial 40 m^2 of public open space per person, it tends to be of poor quality and is often not well utilized

Finally, readers can assess a resident's access to nature by the presence (or otherwise) of a substantial urban forest. The importance of the urban forest to livability in Perth is evidenced by the fact that 77% of respondents to the "The Housing We'd Choose" study considered that the "presence of trees" is an important dwelling attribute (Curtin University and Hames Sharley 2013).

Despite the importance of trees to supporting livability in Perth, one key effect of background infill—in combination with the increasing size of new houses, a lack of tree protection regulation and some residents' "arbor-phobia"—has been a decline in urban forest cover in Perth's inner and middle-ring suburbs (Fig. 2.8). Examples of the effects of background infill development on urban forest cover can be found in the inner and middle-ring local government areas of South Perth, Bayswater, Stirling and Canning (all of which have high infill development targets), with the percentage of total vegetation cleared between 2001 and 2004 amounting to 13%, 12%, 11% and 11%, respectively (McManus 2010). One reason that background infill development has resulted in the removal of urban forest cover is that trees are

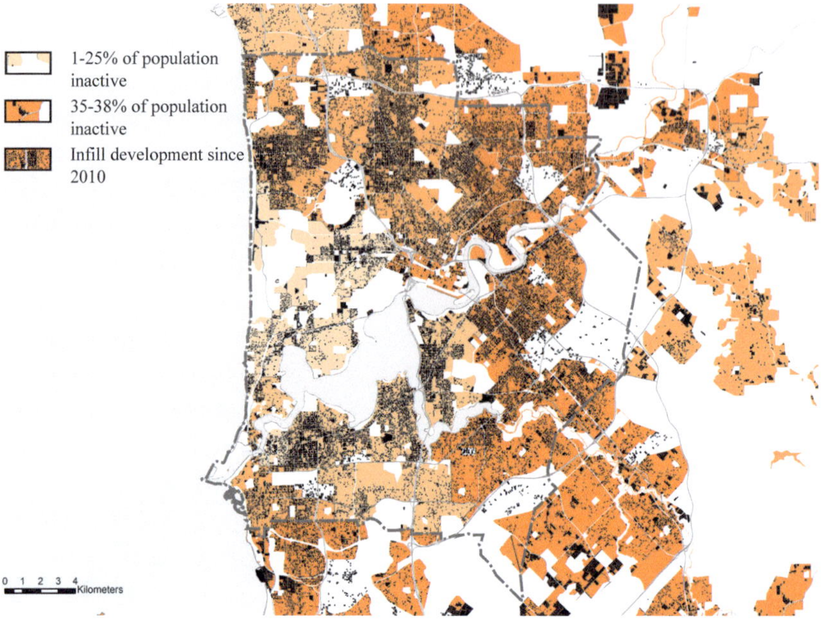

Fig. 2.7 Background infill and levels of physical inactivity: Physical inactivity reaches elevated levels in many areas which have experienced significant background infill. Generally, mediocre POS reserves and a lack of public transport and private open space at least partly explain this

often "treated as trimmings to the designed urban environment and are afforded little or no protection against the exigency of meeting development aspirations" (Brunner and Cozens 2013) (Figs. 2.9 and 2.10).

The clearing of this substantial urban forest has several potential ramifications for the provision of ecosystem services and, in turn, urban livability. In short, the urban forest provides key environmental benefits, including reducing air pollution, sequestering greenhouse gasses (Brunner and Cozens 2013), filtering and cleaning stormwater, minimizing and mitigating urban heat islands (increasingly important in the age of climate change), ameliorating the local climate and supporting biodiversity (Department of Planning and Western Australian Planning Commission 2015).

2.5.5 *Access to Retail Destinations*

Perth's residents regard "easy access" to retail destinations such as local shops, as a crucial dwelling attribute. Indeed, 73% of respondents to "The Housing We'd Choose" survey thought this was an important dwelling attribute (Curtin University and Hames Sharley 2013). So, to what degree is background infill delivering retail assets in Perth? Retail destinations include District Centres, Neighbourhood Centres

Fig. 2.8 Background infill and urban forests: One key effect of background infill has been a decline in urban forest cover in Perth's inner and middle-ring suburbs

Fig. 2.9 Urban forest prior to background infill: A typical suburban area and mature urban forest prior to background infill occurring

Fig. 2.10 Urban forest after background infill: A typical suburban area post background infill occurring. The loss of urban forest cover is partly because trees are afforded little or no protection against meeting development aspirations

and Local Centres, as defined in Perth's Activity Centre policy.[7] When we plotted these centre types against concentrations of background infill, we found a substantial proportion of residential areas that are devoid of such centres (Fig. 2.11). We attribute this to the dispersed nature of background infill, which does not result in the densely populated urban precincts required to support significant commercial assets. As Debra Goostrey, Ex-Chief Executive of the Western Australian Urban Development Industry Association explains:

> When you get to (infill development) precincts, you can suddenly justify a small bar and a restaurant, you have to have enough people in and around that area… density brings with it the coffee shop effect, that great vibrancy that comes through. [But] when you've got small little bits of density, you don't get the coffee shop effect… (In Moodie and Trigger 2015).

In quantitative terms, while the Activity Centre policy defines 6,250 residents per km^2 as a desirable density to support Neighbourhood Centres (based on R25

[7]This policy defines District Centres as "servicing the daily and weekly needs of residents" and their walkable catchment is considered to be 400 m. Typical retail types include department stores, supermarkets and some specialty shops. These are supplemented by smaller scale Neighbourhood Centres that are intended to include a small supermarket, personal services (such as a hairdresser) and convenience shops. Planners typically regard their walkable catchment as 200 m. Finally, Local Centres are defined as any shop with a floor space of less than 1,500 m^2, and usually consist of a "corner deli" and a newsagent (State of Western Australia 2005).

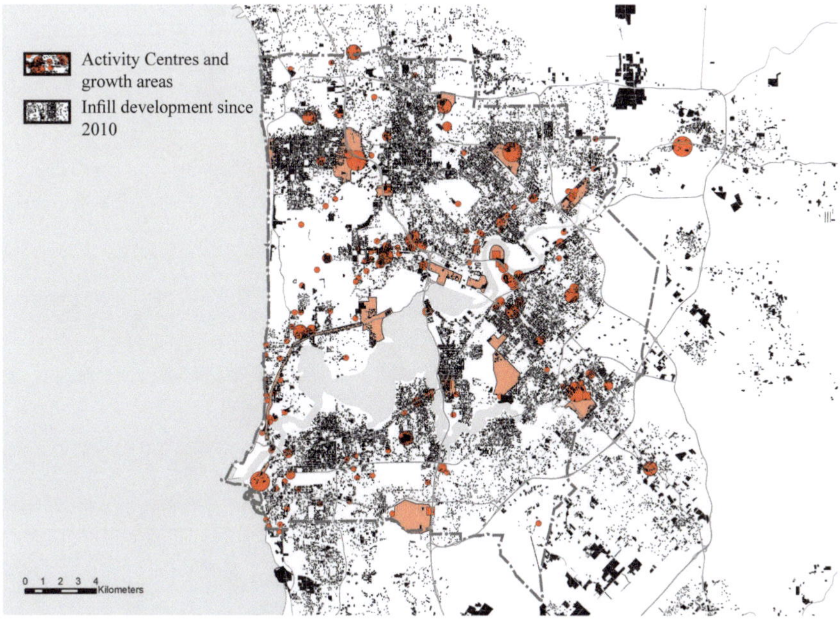

Fig. 2.11 Background infill and Activity Centres: Most background infill that is occurring is not within a walkable catchment of District Centres, Neighbourhood Centres or Local Centres

gross density[8]) even with substantial background infill many Perth local government areas (such as the City of Stirling) only achieve a gross residential density of 2,100 people per km^2 (Australian Bureau of Statistics 2012). Thus, while governments and developers often promote urban densification with images of café culture and the convenience of local shops, it is not clear that the planning rules that perpetuate background infill are delivering this lifestyle.

2.5.6 Accessibility to Public Transport

Both the Economist Intelligence Unit and the Western Australian state government define a vital characteristic of urban livability as being accessibility to public transport (Western Australian Department of Planning 2010; The Economist Intelligence Unit Limited 2012). Despite this, Perth remains a stubbornly car-centric city. So, to what degree are areas of background infill well serviced by public transport in Perth?

Perth's heavy rail system, the backbone of Perth's public transport system, consists of five major rail lines radiating from the city centre. Much of the background infill development that is occurring in middle-ring suburbs is in the substantial areas between the radiating rail lines. Mapping of Perth's existing train stations in combi-

[8]R25 equates to 25 dwellings per ha.

Fig. 2.12 Background infill and train stations: Mapping of Perth's existing train stations in combination with background infill development reveals that little of the infill is within a short walk of train stations

nation with background infill development that has occurred to date reveals that little of the background infill is within a short walk of train stations—in this map shown as an 800 m or 10-minute walk (Fig. 2.12).

Due to this situation, public transport users in these background infill areas rely on Perth's bus system that either feed into the rail system or access the city centre directly. While significantly more of the areas of background infill are serviced by bus routes than trains, the bus system tends to "emphasize coverage" (i.e. short walking distances) at the cost of the "frequency and legibility of the service" (Mees and Dodson 2011) (Fig. 2.13). Mapping of the proportion of people who drive or carpool to work reveals a high vehicular dependency in suburbs with a large amount of background infill, particularly in the northern most section of the middle-ring suburbs (Fig. 2.14).

2.6 Conclusion

In its most bold promise, Peter Calthorpe believed TOD would help to "redefine the American Dream" (Carlton 2009). However, short of the imposition of "draconian planning regimes" (Berger et al. 2017), it is unlikely that most urban TOD devel-

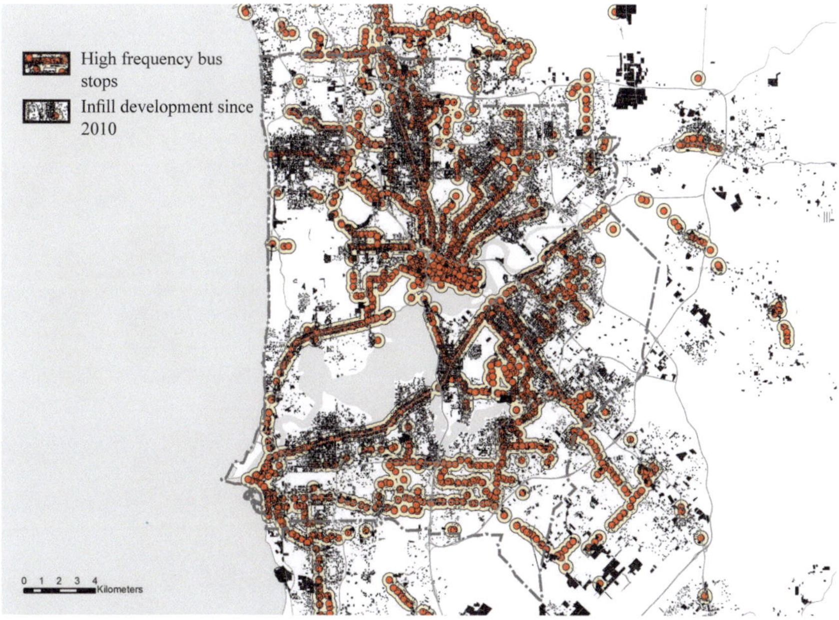

Fig. 2.13 Background infill and bus stops: Bus routes service significantly more the areas of background infill than trains; however, the bus system tends to emphasize coverage at the cost of the frequency and legibility of the service

opment will be achieved. As this chapter has explored, there exists a "divergence between the compact city imagined in metropolitan plans and what is occurring on the ground in Australian cities" (Gray et al. 2010). Despite the application of TOD policy in Perth, Melbourne, Sydney and South East Queensland over many years, implementation has proven a challenge (Kelly and Donegan 2015; Goodman and Moloney 2004; Burton 2017; Goodman 2017; Randolph et al. 2017). This is not a uniquely Australian issue either. Such challenges to TOD are also being experienced in North America in their first generation of "transit towns" (Dittmar et al. 2004). While Calthorpe sought to "redefine the American Dream", we believe most suburban dwellers in the Australian cities are yet to fall under the spell of this dream. This is important because for urban infill to become a viable model, we require an enticing new dream that can compete against the enduring popularity of suburban living.

With Australia's population predicted to triple by 2100 (Australian Bureau of Statistics 2013), the need to deliver high performing infill development is unlikely to ease. While this population growth represents an opportunity for creativity, if policymakers and planners handle it poorly, it could be calamitous. As Brendan Gleeson reminds us—our cities must become the urban "lifeboats" that enable us to "sail through the coming storms of resource shortages and climate change" (Gleeson 2010). In short, delivering infill development in a manner that improves urban

Fig. 2.14 Background infill and car dependency: Mapping of the proportion of people who drive to work reveals a high vehicular dependency in suburbs with a large amount of background infill, particularly in the northernmost section of the middle-ring suburbs

livability will be one of the challenges that defines the viability of Australian cities in this century. In the following chapter, we discuss the potential of GOD to address this emerging situation.

References

Acott K (2019) Is public transport off the rails? The West Australian. https://thewest.com.au/news/wa/is-public-transport-off-the-rails-ng-b881083423z. Accessed 29 Jan 2019

Arvola A, Pennanen K (2014) Understanding residents' attitudes towards infill development at Finnish urban suburbs. Paper presented at the world SB14 Barcelona, Barcelona, 30 September

Australian Bureau of Statistics (2012) 3218.0—regional population growth, Australia, 2011–12. Australian Bureau of Statistics. http://www.abs.gov.au/ausstats/abs@.nsf/mf/3218.0/. Accessed 26 June 2013

Australian Bureau of Statistics (2013) Population projections, Australia, 2012 to 2101. Australian Bureau of Statistics. http://www.abs.gov.au/ausstats/abs@.nsf/Lookup/3222.0main+features52012%20(base)%20to%202101. Accessed 14 Jan 2014

Australian Bureau of Statistics (2017) Apartment living. Australian Bureau of Statistics. https://www.abs.gov.au/ausstats/abs@.nsf/Lookup/by%20Subject/2071.0~2016~Main%20Features~Apartment%20Living~20. Accessed 12 June 2019

Berger A, Kotkin J, Guzman C (2017) Introduction. In: Berger A, Kotkin J, Guzman C (eds) Infinite suburbia. MIT, Boston, pp 10–23

Bolleter J (2015) Scavenging the suburbs: auditing Perth for 1 million infill dwellings. University of Western Australia Publishing, Perth
Bolleter J, Weller R (2013) Made in Australia: the future of Australian cities. University of Western Australia Publishing, Perth
Bruegmann R (2017) The anti-suburban crusade. In: Berger A, Kotkin J, Guzman C (eds) Infinite suburbia. MIT, Boston, pp 26–37
Brunner J, Cozens P (2013) 'Where have all the trees gone?' Urban consolidation and the demise of urban vegetation: a case study from Western Australia. Plan Pract Res 28(2):231–255
Burton P (2017) South East Queensland: change and continuity in planning. In: Hamnett S, Freestone R (eds) Planning metropolitan Australia. Routledge, London, pp 156–177
Carlton I (2009) Histories of transit-oriented development: perspectives on the development of the TOD concept. Institute of Urban & Regional Development, Berkeley
Centre for the Built Environment and Health (2013) Public Open Space (POS) Geographic Information System (GIS) layer. University of Western Australia. http://researchdata.ands.org.au/public-open-space-pos-geographic-information-system-gis-layer. Accessed 11 June 2013
Community Representative (2018) Interview
Congress of New Urbanism (2016) The charter of the New Urbanism. Congress of New Urbanism. https://www.cnu.org/who-we-are/charter-new-urbanism. Accessed 15 Aug 2016
Curtin University, Hames Sharley (2013) The housing we'd choose: a study for Perth and Peel. Department of Housing, Department of Planning, Perth
Curtis C (2010) The Network City metropolitan planning strategy: a paradigm shift for sustainable transport? In: Alexander I, Greive S, Hedgcock D (eds) Planning perspectives from Western Australia: a reader in theory and practice. Fremantle Press, Perth, pp 258–273
Curtis C (2012) Transitioning to transit-oriented development: the case of Perth, Western Australia. Urban Policy Res 30(3):275–292
Davis G, Harford-Mills G (2016) Examining 60 years of strategic planning in metropolitan Perth and Peel. Committee for Perth, Perth
Department of Infrastructure and Planning (2009) South East Queensland regional plan 2009–2031. Department of Infrastructure and Planning, Brisbane
Department of Infrastructure LGaP (2017) Shaping SEQ: South East Queensland regional plan 2017. Queensland Government, Brisbane
Department of Planning, Western Australian Planning Commission (2015) Draft Perth and Peel @3.5 million. Western Australian Planning Commission, Perth
Department of Planning Lands and Heritage (2019) Urban growth monitor: Perth metropolitan, Peel and Greater Bunbury regions. Western Australian Planning Commission, Perth
Department of Planning TaI (2017) The 30-year plan for greater Adelaide: 2017 update. Department of Planning, Transport and Infrastructure, Adelaide
Development Institute Representative (2018) Interview
Dittmar H, Belzer D, Autler G (2004) An introduction to transit-oriented development. In: Dittmar H, Ohland G (eds) New Transit Town: best practices in transit-oriented development. Island Press, Washington
Dodson J (2010) In the wrong place at the wrong time? Assessing some planning, transport and housing market limits to urban consolidation policies. Urban Policy Res 28(4):487–504
Dovey K, Woodcock I (2014) Intensifying Melbourne: transit-oriented urban design for resilient urban futures. Melbourne School of Design, The University of Melbourne, Melbourne
Elliot R (2017) Australia's misplaced war on the Australian dream. In: Berger A, Kotkin J (eds) Infinite suburbia. MIT, Boston, pp 104–113
Falconer R, Babb C, Olaru D (2016) Cities as systems: node and place conflict across a rail transit network. In: Biermann S, Olaru D, Paul V (eds) Planning boomtown and beyond. UWA Publishing, Perth, pp 460–489
Gillen M (2006) The challenge of attaining a sustainable urban morphology for South East Queensland. Plan Pract Res 21(3):291–308
Gleeson B (2010) Lifeboat cities. UNSW Press, Sydney

Gleeson B, Dodson J, Spiller M (2010) Metropolitan governance for the Australian city: the case for reform. Issues Pap 12(1):1–26
Goodman R (2017) Melbourne: growing pains for the liveable city. In: Hamnett S, Freestone R (eds) Planning metropolitan Australia. Routledge, pp 59–83
Goodman R, Moloney S (2004) Activity centre planning in Melbourne revisited. Aust Plan 41(2)
Government of Western Australia (2010) State planning policy 4.2: activity centres for Perth and Peel. Government of Western Australia, Perth
Gray R, Gleeson B, Burke M (2010) Urban consolidation, household greenhouse emissions and the role of planning. Urban Policy Res 28(3):335–346
Hagan S (2017) Metabolic suburbs or the virtue of low densities. In: Berger A, Kotkin J, Guzman C (eds) Infinite suburbia. MIT, Boston, pp 468–477
Holling C, Haslam McKenzie F (2010) Integrated transit-oriented-development: is it appropriate for Perth? In: Alexander I, Greive S, Hedgcock D (eds) Planning perspectives from Western Australia: a reader in theory and practice. Fremantle Press, Perth, pp 274–288
Hurley J, Taylor E, Dodson J (2017) Why has urban consolidation been so difficult. In: Sipe N, Vella K (eds) The Routledge handbook of Australian urban and regional planning, New York, pp 123–135
Infrastructure Australia (2018) Future cities: planning for our growing population
Kelly J-F, Donegan P (2015) City limits: why Australian cities are broken and how we can fix them. Melbourne University Press, Melbourne
Local Government Planner (2018) Interview
Maginn P, Foley N (2017) Perth: from 'Large Provincial City' to 'Globalizing City'. In: Hamnett S, Freestone R (eds) Planning metropolitan Australia. Routledge, London, pp 124–147
McManus P (2010) Planning with and for trees in Perth: yesterday, today and tomorrow. In: Alexander I, Greive S, Hedgcock D (eds) Planning perspectives from Western Australia: a reader in theory and practice. Fremantle Press, Perth, pp 340–353
Mees P, Dodson J (2011) Public transport network planning in Australia: assessing current practice in Australia's five largest cities. Griffith University Urban Research Program, no 34, pp 1–28
Moodie C, Trigger R (2015) Perth infill backlash: suburbs fighting high-density development. ABC News. http://www.abc.net.au/news/2015-06-08/perth-infill-backlash-suburbs-fighting-high-density-development/6521460?WT.ac=localnews_perth. Accessed 09 June 2015
Murphy P (2012) The metropolis. In: Maginn P, Thompson S (eds) Planning Australia: an overview of urban and regional planning. Cambridge University Press, Melbourne, pp 155–179
Newman P (2018) Why trackless trams are ready to replace light rail. The conversation. https://theconversation.com/why-trackless-trams-are-ready-to-replace-light-rail-103690. Accessed 30 Dec 2018
OECD (2012) Compact city policies: a comparative assessment. OECD Green Growth Studies
Parliamentarian (2018) Interview
PHIDU Torrens University Australia Social Health Atlases. Torrens University Australia. http://phidu.torrens.edu.au/social-health-atlases#estCBaZ1koZPR8k3.97. Accessed 14 Aug 2016
Randolph B, Freestone R, Bunker R (2017) Sydney: growth, globalization and governance. In: Hamnett S, Freestone R (eds) Planning metropolitan Australia. Routledge, London, pp 84–108
Rice J (2016) There goes the neighbourhood? Or saving the world? Community views about transit oriented development. In: Curtis C, Renne JL, Bertolini L (eds) Transit oriented development: making it happen. Routledge, New York, pp 191–204
Rowley S, Phibbs P (2012) Delivering diverse and affordable housing on infill development sites, vol 193. Australian Housing and Urban Research Institute, Melbourne
Sarkissian W (2013) Wendy Sarkissian on NIMBYism, community resistance and housing density. The Fifth Estate. http://www.thefifthestate.com.au/spinifex/nimbyism-community-resistance-and-housing-density/45397. Accessed 17 Dec 2015
Seddon G (1994) The Australian back yard. In: Craven I (ed) Australian popular culture. Cambridge University Press, Cambridge, pp 22–35

South East Queensland Regional Organisation of Councils (2005) South East Queensland regional plan 2005–2026. Office of Urban Management, Queensland Department of Local Government, Brisbane

State of Western Australia (2005) State planning policy 4.2: activity centres for Perth and Peel. State of Western Australia, Perth

State of Western Australia (2010) State planning policy 3.1: residential design codes. State of Western Australia, Perth

Syme G, Fenton M, Coakes S (2001) Lot size, garden satisfaction and local park and wetland visitation. Landsc Urban Plan 56:161–170

The Economist Intelligence Unit Limited (2012) Best cities ranking and report: a special report from the Economist Intelligence Unit. The Economist, London

Troy P (2004) Saving our cities with suburbs. In: Schultz J (ed) Griffith review: dreams of land. Griffith University, Brisbane

Victoria State Government (2017) Plan Melbourne 2017–2050. Victoria State Government, Melbourne

Weller R (2009) Boomtown 2050. University of Western Australia Press, Perth

Western Australian Department of Planning (2010) Directions 2031 and beyond: metropolitan planning beyond the horizon. Department of Planning, Perth

Chapter 3
Why GOD? The Benefits of Greenspace-Oriented Development

Abstract In this chapter, we set out a strategy for urban densification, which we name Greenspace-Oriented Development (GOD). While Transit-Oriented Development (TOD) co-locates urban densification with public transport hubs, GOD co-locates urban densification with significant, upgraded public green spaces (such as parks) that are relatively well served by public transport. The upgraded parks, and their connecting streetscapes, operate as a multifunctional, communal "backyard" for residents living in a surrounding higher-density urban precinct. Through GOD, we propose to weave together the positive aspects of suburbia (i.e. access to open space and nature) with those of good quality medium-density urban infill (i.e. access to public transport, facilities and good urban design). In this chapter, we describe GOD, its key principles, benefits and the settings to which it is most suited.

Keywords Ecosystem services · Benefits of urban green spaces · Greenspace-Oriented Development · Transit-Oriented Development · Urban biodiversity · Urban consolidation · Urban design · Urban green space · Urban infill · Urban parks · Public open space

3.1 An Overview of GOD

As set out in the previous chapter, the principles of TOD are well established, yet it confronts many challenges to implementation. We believe that we need a complementary strategy for achieving infill development, and that Greenspace-Oriented Development (GOD) is this viable strategy. This is because GOD is eminently implementable but also offers a host of environmental and human health and well-being benefits.

The GOD approach acts on three main components of the suburban landscape: (1) target parks, (2) surrounding urban precincts and (3) connecting streetscapes, and in simple terms correlates urban densification with significant, upgraded public parks. We propose that these parks operate as a multifunctional, communal "backyard" for residents living in an adjacent higher-density urban precinct. The parks should be of a significant size, at least greater than 1 ha, and should offer reasonable connectivity to public transport, for example, a 5-minute cycle or a 15–20-minute

J. Bolleter and C. E. Ramalho, *Greenspace-Oriented Development*,
SpringerBriefs in Geography, https://doi.org/10.1007/978-3-030-29601-8_3

walk (approximately 1,600 m) to train stations or rapid bus transit. Upgraded connecting streetscapes (e.g. between the park and nearby transport hubs, but also shops, facilities, community centres and schools) provide connectivity and deliver further communal green areas, improve visual quality and help activating the neighbourhood. The higher-density urban precinct consists of medium-rise apartments (adjacent to the park) through to low-rise apartments and terrace house dwellings located within a 400 m radius of the park (Fig. 3.1).

We believe that a GOD approach can promote urban densification in suburban settings, from older inner-ring suburbs through to middle-ring (greyfield) and outer ring (greenfield) suburbs. In this and the next chapter, we focus on the potential of GOD for Australia's greyfield, middle-ring suburbs. These were developed between the 1950s and 1970s (Newton et al. 2011) and typically comprise underutilized and outdated property assets, with a residential building stock that is up for replacement (Newton 2010). Because of their relative proximity to city centres, greyfield suburbs have high redevelopment and densification potential, and consequently high infill targets.

While Australia's greyfield suburbs contain a reasonable number of parks, many of these are under-designed, offer minimal amenity and are typically underutilized (Figs. 3.2 and 3.3). Indeed, local governments often swathe this ubiquitous component of the greyfield landscape with irrigated lawn and a scattering of trees. Moreover, the design of greyfield parks caters mainly for organized active team sports, over

Fig. 3.1 Greenspace-Oriented Development: GOD co-locates urban densification with upgraded public green spaces that are relatively well served by public transport

Fig. 3.2 Time-lapse photography of a greyfield park in Perth: The design of greyfield parks caters for organized active team sports, over many other community and ecosystem services. *Source* Julian Bolleter

Fig. 3.3 Time-lapse photography of a greyfield park in Perth: Many greyfield parks are under-designed, offer minimal amenity, and are typically underutilized. Time-lapse photography of a greyfield park in Perth reveals a sparsity of use. *Source* Julian Bolleter

many other community and ecosystem services (e.g. passive recreation and wildlife habitat). Indeed, the main facilities provided by local governments include those designed to enable team sports, such as clubrooms, goal posts, basketball hoops and cricket pitches (Byrne and Sipe 2010, p. 6).

The focus on active recreation in greyfield parks is the result of the "recreation movement" (Byrne and Sipe 2010, p. 6), which was prevalent in the mid-twentieth century when Australia's greyfields expanded significantly. The recreation movement proposed that "what were needed most were opportunities for citizens to exercise, to strengthen and discipline bodies, to temper immoral impulses and to give people a place to vent frustrations and escape from urban life" (Byrne and Sipe 2010, p. 6). This movement saw a shift from highly wrought landscaped parks that were dominant in the nineteenth and early twentieth centuries, to highly simplified parks catering for active recreation (Byrne and Sipe 2010, p. 6). At that time, most Australian capital cities still retained large tracts of remnant vegetation surrounding them and, therefore, these open, sports-oriented green spaces of the "recreation movement" offered a "clean" green slate in the then "urban–wildland" interface. Nonetheless, as cities expanded, greyfield suburbs got further and further away from the natural areas on the city's periphery, and residents increasingly experienced a lack of access to nature and good quality, multifunctional green spaces.

3.2 Why Focus Densification Around Upgraded Parks?

We have based the association between urban densification and the provision of good quality green spaces in GOD on three key principles. First, green spaces can provide a range of human health and well-being, ecological and economic benefits, and "compensate" residents living in higher-density settings for a relative lack of private green space (Haaland and van den Bosch 2015; Chiesura 2004) (Fig. 3.4). Second, well-designed, higher-density urban precincts surrounding parks can offer important benefits to the utility of the parks themselves. This includes increased local rates and taxes that local governments can direct towards park upgrades and maintenance, and more people to activate the park and increase its safety (Udell et al. 2014). Finally, by being able to promote the socio-economic rejuvenation of the nearby urban areas (e.g. Ryu and Kwon 2016; LaFarge 2014), namely by increasing their property values (Panduro and Veie 2013; Brander and Koetse 2011; Crompton 2005), green spaces can foster urban redevelopment and densification (Mell 2009; Newton et al. 2011). We discuss this process in detail in the next chapter.

Several complexities and nuances are, however, present in this premise. First, the benefits provided by green spaces depend on their attributes and, therefore, how well they cater for different functions, users and their needs (Giles-Corti et al. 2012; Francis et al. 2012; Sugiyama et al. 2015). A typical Australian greyfield park, with

Fig. 3.4 The benefits of GOD: Upgraded parks can provide a range of human health and well-being, ecological and economic benefits. Moreover, well-designed, higher-density urban precincts surrounding parks can offer important benefits to the utility of the parks themselves

its openness and paucity of social and ecological features, is likely to provide a limited range of benefits to its surrounding neighbourhood, most of which relate to formal or informal pop-up sports events. Second, not all types of green spaces positively affect property values. While parks and lakes are associated with large price premiums (Panduro and Veie 2013), especially in high-density settings (Brander and Koetse 2011), sports fields may not have a significant effect (Panduro and Veie 2013; Pandit et al. 2014). Third, and finally, the upgrade of public green spaces (and the making of suburbia more livable and attractive) can have the paradoxical result of leading to ecological or environmental gentrification (Dooling 2009; Checker 2011). This can reduce housing opportunities for low-income residents and affect the commercial/retail infrastructure that supports them (Wolch et al. 2014; Haase et al. 2017). Even small or ecologically oriented urban renewal projects have unintendedly displaced low-income residents (see Wolch et al. 2014; Haase et al. 2017). Awareness of these complexities does not diminish the value of our proposed GOD but instead provides guidance to how local governments, amongst others, should implement it (Chap. 4).

3.3 What Are the Benefits of Green Spaces to Residents in Higher-Density Settings?

Increased density through infill development often means the loss of green spaces and canopy cover, especially in private residential areas, vacant land and areas of undeveloped remnant vegetation (Lin et al. 2015; Hall 2010; Brunner and Cozens 2013; Pauleit et al. 2005). Increased density also means catering for a larger and more diverse population. For this reason, enhancing the provision of green space benefits to residents in higher-density settings requires careful attention to green space quantity, as well as quality, design and accessibility (Haaland and van den Bosch 2015; Byrne and Sipe 2010; Byrne et al. 2010).

In the following paragraphs, we explore the key benefits of green spaces in the context of higher density—for people, biodiversity and the local environment. We focus on parks, as we have oriented GOD towards densification around them. When appropriate, we also highlight the role of street trees, streetscape plantings and other small green spaces (e.g. green roofs and facades), as these can be used in the upgrade of connecting streetscapes, and in the design of the higher-density urban precincts themselves (Figs. 3.5, 3.6, 3.7, 3.8 and 3.9).

3.3.1 Human Health and Well-Being

Contact with nature in green spaces has a crucial role in people's quality of life, influencing health and well-being through four key pathways (Hartig et al. 2014): first,

Fig. 3.5 The benefits of green spaces in higher-density settings: Urban green spaces can provide a range of human well-being, social, ecological and economic benefits, and "compensate" residents living in higher-density settings for the lack of private green space. Urban Park in Melbourne, Australia. *Source* Cristina E. Ramalho

by promoting physical activity (e.g. recreational walking and outdoor play); second, reducing exposure to stress factors and providing an environment for physiological and mental recovery that delivers coping resources to deal with life stressors (Myers 2019); third, promoting social interaction and sense of community; and fourth, providing a healthy, comfortable urban environment (e.g. better air quality and thermal comfort). Overall, these pathways lead to multiple health and well-being benefits that play out across an individual's life span (Astell-Burt et al. 2014). In some cases, people can benefit from nature even without consciously engaging with it (e.g. a natural view from a window; Ulrich 1984; Cox et al. 2019).

While increased urban density can per se promote walkability (in contrast with low-density, car-dependent neighbourhoods; Udell et al. 2014; Giles-Corti et al. 2012), recreational walking and physical activity in green spaces helps to combat sedentary lifestyles and is associated with a reduction in obesity, heart disease, several types of cancer and with extended life spans (Pereira et al. 2013; Bell et al. 2008; Yelenik and Levine 2011). Not surprisingly, physical activity in green settings, such as a walk or a run in the park, is more restorative than conducting the same activity in the built environment (Marselle et al. 2013). Good quality parks and small green spaces in residential precincts also offer residents living at higher densities many of the recreational benefits of traditional suburban gardens. These include allowing residents to pursue a variety of hobbies related with the natural environment, to have

Fig. 3.6 The benefits of green spaces in higher-density settings: Aside from the main urban parks, GOD precincts and surrounding connecting streetscapes can use a myriad of other small green spaces in their design. These small green spaces provide further socio-ecological benefits and contribute to the visual quality of the residential areas. *Source* Cristina E. Ramalho

dogs (Coleman et al. 2008), to cultivate fruit and vegetables and to allow a degree of personal expression through gardening (Troy 2004; Seddon 1994).

Active and passive recreation in—or even indirect exposure to—green spaces also reduces stress and the psychological toll of urban living (Tyrväinen et al. 2014; Peschardt and Stigsdotter 2013; Bratman et al. 2015; De Vries et al. 2013), improves mental health (Francis et al. 2012), attention restoration (Nordh et al. 2009) and plays a crucial role in the cognitive development of children (Dadvand et al. 2015). Indeed, urban parks and other green spaces provide children with unique opportunities for risk-taking, discovery, creativity, mastery and control, which strengthens sense of self, inspires basic emotional states and enhances psychological restoration (Bowler et al. 2010). A study in Massachusetts showed that greenness within a 2 km radius of schools was positively associated with children's academic performance during springtime (Wu et al. 2014). Frequent contact with nature also reduces the symptoms of Attention-Deficit Hyperactivity Disorder in children (Kuo and Faber Taylor 2004). Within built precincts, different greening solutions, such as green roofs, can also have positive effects on attention restoration and stress reduction (e.g. Lee et al. 2015).

Fig. 3.7 The benefits of green spaces in higher-density settings: Community vegetable gardens can be embedded in the GOD parks, as well as within the surrounding higher-density urban precincts. *Source* Cristina E. Ramalho

Parks and other green spaces are also important to create and reinforce social cohesion and sense of community. Indeed, good quality parks provide a forum for socializing with friends and neighbours, and thus improve social ties (Kaźmierczak 2013). A large study in the Netherlands showed that people with more parks and other large green spaces within 1 km of their homes felt healthier, less lonely and more socially supported than those without such amenities nearby. This pattern was stronger in high-density areas and for people with a low income, as well as children and the elderly who, as a consequence of limited mobility, rely more on their neighbourhood to support their needs (Maas et al. 2009). A related study found that people with more good quality streetscape vegetation felt their neighbourhood was calmer and more cohesive, and this was related to improved self-reported health (De Vries et al. 2013). In Chicago, parks were found to indirectly mitigate stress by fostering social support, an effect that other urban vegetation did not have (Fan et al. 2011).

As to the effect of biodiversity on human well-being, Fuller et al. (2007) and Dallimer et al. (2012) showed that greater perceived richness of plants and birds in urban parks, as well as greater habitat diversity, were associated with increased psychological benefits to park users. More recently, Carrus et al. (2015) and Marselle et al. (2016) showed that this association is mediated by the perceived restorative quality of parks with those characteristics. In other words, perceived biodiversity—and naturalness (Marselle et al. 2016)—enhances people's perceived restorative effect

Fig. 3.8 The benefits of green spaces in higher-density settings: Large mature street trees provide various socio-ecological benefits, including shade, aesthetics and sense of place, and they can form the centrepiece of small green spaces and informal seating and/or resting areas (White Gum Valley, Fremantle, Australia). *Source* Cristina E. Ramalho

of green spaces, which is associated with greater levels of well-being. In Berlin, Palliwoda et al. (2017) concluded that a considerable proportion of leisure activities undertaken by park users were linked to biodiversity, in particular, individual plant species, with some species sought after for consumption and decoration, while others for simple *in situ* observation and experience.

Urban biodiversity also allows people to meaningfully engage with nature and learn about the natural world. As cities are more spread out than ever before, people are likely to primarily experience nature in the places where they live and work (Miller and Hobbs 2002). Connection with nature in urban areas is thus key to prevent the extinction of nature's experience (Miller 2005). Furthermore, it fosters environmental stewardship in the places where people live, but also beyond cities' boundaries (Dunn et al. 2006). Connection with urban nature in childhood underpins lifetime commitment towards environmental protection, as stated by many professionals and volunteers working in this field (van den Born et al. 2018). Furthermore, it adds meaningfulness and satisfaction to people's lives (Chan et al. 2016).

Fig. 3.9 The benefits of decent quality green spaces: Green walls and green facades represent another greening solution to be potentially used in GOD precincts and connecting streetscapes. These green spaces are especially useful when the ground level is largely sealed (green wall in Paris, France). *Source* Cristina E. Ramalho

3.3.2 Biodiversity Conservation

Urban green spaces are not only important for people but also to the other species we share our cities with. There are more threatened species in Australian cities per unit of area than elsewhere in the country (Ives et al. 2016), a reality that mirrors the US (Schwartz et al. 2002). This reflects the negative impact that urbanization has on biodiversity—driving species to endangerment—but it also reflects a different, emerging reality. Some threatened species are attracted to the urban environment because of the multiple resources available there (e.g. additional water and food, built-up structures that mimic nesting areas; Ives et al. 2016). Parks containing remnant vegetation or semi-natural habitats are crucial for biodiversity conservation (Ramalho et al. 2014; e.g. Koh and Sodhi 2004), even if they are only small areas (Kendal et al. 2017). Parks with quality native plantings, including understorey vegetation (shrubs and herbaceous plants), can also provide important refugia for biodiversity, and stepping stones that facilitate species movement through the urban landscape (Threlfall et al. 2015, 2017). Other types of green spaces, including street trees, streetscape plantings and green roofs, can also play a key role in biodiversity conservation, namely by providing habitat and enhancing ecological connectivity (e.g. Mullaney et al. 2015).

3.3.3 Services to the Local Environment

Urban green spaces can improve the quality of the local environment through the provision of several ecosystem services. Reasonable canopy cover in parks, streetscapes or even residential precincts can cool the microclimate and improve thermal comfort (Coutts et al. 2016). While trees can cool down the microclimate through evapotranspiration, this is generally a small effect compared to that provided by their shading, which can lead to substantial localized reductions in surface temperature (Coutts et al. 2016). The role of trees in microclimatic regulation and thermal comfort is becoming particularly relevant in a climate change context, given that the combination of global warming and Urban Heat Island is leading to increased temperatures in cities worldwide (Coutts et al. 2010). Such temperatures can pose a serious threat to children and the elderly during extreme heat events (Kovats and Hajat 2008). Moreover, trees and understorey vegetation in parks, streetscapes and residential areas can provide several other ecosystem services, reducing noise from surrounding streets and roads, filtering air pollutants, and storing carbon (Bolund and Hunhammar 1999; Tzoulas et al. 2007). Soils and vegetation also intercept rainfall, decreasing runoff into stormwater channels and increasing water infiltration into soils (Nouri et al. 2013; Ossola et al. 2015). Within built precincts, green solutions and technologies can improve indoor air quality and thermal comfort (Xing et al. 2017; Balaban and de Oliveira 2017).

3.4 The Benefits of Well-Designed Higher-Density Urban Precincts for Co-located Parks

Appropriately designed higher-density urban precincts surrounding public parks can offer important benefits to the utility of the parks themselves. Indeed, an increased population paying local rates and taxes means greater income, which local governments can direct towards park upgrades and maintenance. An increased population also means more people in the vicinity to activate the park and increase its sense of enjoyment and safety through passive surveillance or "eyes on the street" (Udell et al. 2014).

Additionally, a higher-density urban precinct surrounding a public green space provides the ideal opportunity for experimentation of urban sustainability approaches, including green building technology, urban design solutions and urban living labs (Felson et al. 2013; Voytenko et al. 2016) (Figs. 3.10 and 3.11). Examples of urban design solutions that could improve public parks

Fig. 3.10 GOD precedents: There are multiple examples worldwide that could inspire the planning and design of GOD precincts. The Western Harbour Development in Malmo, Sweden, is one such example. This development provides a model for urban sustainability, with well-designed medium-density urban precincts built using green building technology and employing greening solutions in private space (e.g. green roofs), while surrounding relatively small good quality, attractive public green spaces. *Source* Cristina E. Ramalho

include the collection of communal composting and its use as fertilizer in garden beds and communal vegetable gardens; the harvesting of grey water for irrigation; and the creation of bio-retention gardens for flood mitigation and provision of water features and aquatic habitat (Smith et al. 2009; Felson et al. 2013). Urban living labs extend beyond the experimentation of design and technological solutions. Rather, they entail a form of collective, multi-actor governance that aims to creatively co-generate innovative social solutions for urban living (Voytenko et al. 2016), including the co-design and establishment of places with symbolic meaning (Frantzeskaki et al. 2018) (likely in the upgraded park). Without spaces of meaning and narrative, attempts at infill development "will struggle to be more than a duplicitous rendering in a real-estate brochure" (Barns and Mar 2018). Finally, experimentation of urban sustainability solutions (and their monitoring and testing) engages residents with sustainability goals (Felson et al. 2013; Smith and Billig 2012). It can challenge the public discourse about urban densification by illustrating the benefits and synergies that can be achieved in higher-density settings (Udell et al. 2014). Indeed, upgraded parks provides an excellent opportunity to publicly display such benefits.

Fig. 3.11 GOD precedents: An example of a small urban green space (with a playground to the right) at the centre of a mixed-density residential complex in the Western Harbour Development, Malmo (Sweden). *Source* Cristina E. Ramalho

3.5 Conclusion

We believe that well-designed medium-density urban precincts surrounding a decent quality public green space can offer residents many of the benefits of leafy suburban form with its "green neighborhoods, fresh air, pleasant views and shady gardens" (Troy 2004). Moderately dense urban forms can bring destinations closer together, including local shops and facilities, public transport accessibility, and improved cultural amenity. Public parks, but also street trees and streetscape plantings, not only provide a myriad of social and ecological benefits to urban dwellers, they can also play a crucial role in activating neighbourhoods and create places with meaning and sense of place which people feel attached to. Parks of the GOD approach are ideal places for experimenting with design, technology, social and governance solutions that can help transition urban environments to higher sustainability and livability standards. Given these potential synergies, we believe that GOD provides a viable, promising approach to support sustainable infill in Australian suburbia.

References

Astell-Burt T, Mitchell R, Hartig T (2014) The association between green space and mental health varies across the lifecourse. A longitudinal study. J Epidemiol Community Health 68(6):578–583

Balaban O, de Oliveira JAP (2017) Sustainable buildings for healthier cities: assessing the co-benefits of green buildings in Japan. J Clean Prod 163:S68–S78

Barns S, Mar P (2018) Re-imagining Parramatta: negotiating the arrival of Australia's next great city. Griffith Rev 61:223

Bell JF, Wilson JS, Liu GC (2008) Neighborhood greenness and 2-year changes in body mass index of children and youth. Am J Prev Med 35(6):547–553

Bolund P, Hunhammar S (1999) Ecosystem services in urban areas. Ecol Econ 29(2):293–301. https://doi.org/10.1016/s0921-8009(99)00013-0

Bowler D, Buyung-Ali LM, Knight TM, Pullin AS (2010) A systematic review of evidence for the added benefits to health of exposure to natural environments. BMC Public Health 10(1):456

Brander LM, Koetse MJ (2011) The value of urban open space: meta-analyses of contingent valuation and hedonic pricing results. J Environ Manage 92(10):2763–2773. https://doi.org/10.1016/j.jenvman.2011.06.019

Bratman GN, Daily GC, Levy BJ, Gross JJ (2015) The benefits of nature experience: improved affect and cognition. Landsc Urban Plan 138:41–50. https://doi.org/10.1016/j.landurbplan.2015.02.005

Brunner J, Cozens P (2013) 'Where have all the trees gone?' Urban consolidation and the demise of urban vegetation: a case study from Western Australia. Plann Pract Res 28(2):231–255. https://doi.org/10.1080/02697459.2012.733525

Byrne J, Sipe N (2010) Green and open space planning for urban consolidation—a review of the literature and best practice. In: Urban research program, vol Issues Paper 11. Griffith University, Brisbane, Australia

Byrne J, Sipe N, Searle G (2010) Green around the gills? The challenge of density for urban greenspace planning in SEQ. Aust Plan 47(3):162–177

Carrus G, Scopelliti M, Lafortezza R, Colangelo G, Ferrini F, Salbitano F, Agrimi M, Portoghesi L, Semenzato P, Sanesi G (2015) Go greener, feel better? The positive effects of biodiversity on the well-being of individuals visiting urban and peri-urban green areas. Landsc Urban Plan 134:221–228. https://doi.org/10.1016/j.landurbplan.2014.10.022

Chan KM, Balvanera P, Benessaiah K, Chapman M, Díaz S, Gómez-Baggethun E, Gould R, Hannahs N, Jax K, Klain S (2016) Opinion: why protect nature? Rethinking values and the environment. Proc Natl Acad Sci 113(6):1462–1465

Checker M (2011) Wiped out by the "Greenwave": environmental gentrification and the paradoxical politics of urban sustainability. City Soc 23(2):210–229. https://doi.org/10.1111/j.1548-744X.2011.01063.x

Chiesura A (2004) The role of urban parks for the sustainable city. Landsc Urban Plan 68(1):129–138. https://doi.org/10.1016/j.landurbplan.2003.08.003

Coleman KJ, Rosenberg DE, Conway TL, Sallis JF, Saelens BE, Frank LD, Cain K (2008) Physical activity, weight status, and neighborhood characteristics of dog walkers. Prev Med 47(3):309–312. https://doi.org/10.1016/j.ypmed.2008.05.007

Coutts A, Beringer J, Tapper N (2010) Changing urban climate and CO2 emissions: implications for the development of policies for sustainable cities. Urban Policy Res 28(1):27–47. https://doi.org/10.1080/08111140903437716

Coutts AM, White EC, Tapper NJ, Beringer J, Livesley SJ (2016) Temperature and human thermal comfort effects of street trees across three contrasting street canyon environments. Theoret Appl Climatol 124(1–2):55–68

Cox DTC, Bennie J, Casalegno S, Hudson HL, Anderson K, Gaston KJ (2019) Skewed contributions of individual trees to indirect nature experiences. Landsc Urban Plan 185:28–34. https://doi.org/10.1016/j.landurbplan.2019.01.008

Crompton JL (2005) The impact of parks on property values: empirical evidence from the past two decades in the United States. Manag Leis 10(4):203–218
Dadvand P, Nieuwenhuijsen MJ, Esnaola M, Forns J, Basagaña X, Alvarez-Pedrerol M, Rivas I, López-Vicente M, Pascual MDC, Su J (2015) Green spaces and cognitive development in primary schoolchildren. Proc Natl Acad Sci 112(26):7937–7942
Dallimer M, Irvine KN, Skinner AMJ, Davies ZG, Rouquette JR, Maltby LL, Warren PH, Armsworth PR, Gaston KJ (2012) Biodiversity and the feel-good factor: understanding associations between self-reported human well-being and species richness. Bioscience 62(1):47–55. https://doi.org/10.1525/bio.2012.62.1.9
De Vries S, Van Dillen SM, Groenewegen PP, Spreeuwenberg P (2013) Streetscape greenery and health: stress, social cohesion and physical activity as mediators. Soc Sci Med 94:26–33
Dooling S (2009) Ecological gentrification: a research agenda exploring justice in the city. Int J Urban Reg Res 33(3):621–639
Dunn RR, Gavin MC, Sanchez MC, Solomon JN (2006) The pigeon paradox: dependence of global conservation on urban nature. Conserv Biol 20(6):1814–1816. https://doi.org/10.1111/j.1523-1739.2006.00533.x
Fan Y, Das KV, Chen Q (2011) Neighborhood green, social support, physical activity, and stress: assessing the cumulative impact. Health Place 17(6):1202–1211. https://doi.org/10.1016/j.healthplace.2011.08.008
Felson AJ, Bradford MA, Terway TM (2013) Promoting earth stewardship through urban design experiments. Front Ecol Environ 11(7):362–367
Francis J, Wood LJ, Knuiman M, Giles-Corti B (2012) Quality or quantity? Exploring the relationship between Public Open Space attributes and mental health in Perth, Western Australia. Soc Sci Med 74(10):1570–1577
Frantzeskaki N, van Steenbergen F, Stedman RC (2018) Sense of place and experimentation in urban sustainability transitions: the Resilience Lab in Carnisse, Rotterdam, The Netherlands. Sustain Sci 13:1045–1059
Fuller RA, Irvine KN, Devine-Wright P, Warren PH, Gaston KJ (2007) Psychological benefits of greenspace increase with biodiversity. Biol Let 3:390–394
Giles-Corti B, Ryan K, Foster S (2012) Increasing density in Australia: maximising the health benefits and minimising the harm. National Heart Foundation of Australia, Melbourne
Haaland C, van den Bosch CK (2015) Challenges and strategies for urban green-space planning in cities undergoing densification: a review. Urban For Urban Green 14(4):760–771
Haase D, Kabisch S, Haase A, Andersson E, Banzhaf E, Baró F, Brenck M, Fischer LK, Frantzeskaki N, Kabisch N, Krellenberg K, Kremer P, Kronenberg J, Larondelle N, Mathey J, Pauleit S, Ring I, Rink D, Schwarz N, Wolff M (2017) Greening cities—to be socially inclusive? About the alleged paradox of society and ecology in cities. Habitat Int 64:41–48. https://doi.org/10.1016/j.habitatint.2017.04.005
Hall T (2010) Goodbye to the backyard? The minimisation of private open space in the Australian outer-suburban estate. Urban Policy Res 28(4):411–433
Hartig T, Mitchell R, De Vries S, Frumkin H (2014) Nature and health. Annu Rev Public Health 35:207–228
Ives CD, Lentini PE, Threlfall CG, Ikin K, Shanahan DF, Garrard GE, Bekessy SA, Fuller RA, Mumaw L, Rayner L (2016) Cities are hotspots for threatened species. Glob Ecol Biogeogr 25(1):117–126
Kaźmierczak A (2013) The contribution of local parks to neighbourhood social ties. Landsc Urban Plan 109(1):31–44. https://doi.org/10.1016/j.landurbplan.2012.05.007
Kendal D, Zeeman BJ, Ikin K, Lunt ID, McDonnell MJ, Farrar A, Pearce LM, Morgan JW (2017) The importance of small urban reserves for plant conservation. Biol Cons 213:146–153
Koh LP, Sodhi NS (2004) Importance of reserves, fragments, and parks for butterfly conservation in a tropical urban landscape. Ecol Appl 14(6):1695–1708. https://doi.org/10.1890/03-5269
Kovats RS, Hajat S (2008) Heat stress and public health: a critical review. Annu Rev Public Health 29:41–55

Kuo FE, Faber Taylor A (2004) A potential natural treatment for attention-deficit/hyperactivity disorder: evidence from a national study. Am J Public Health 94(9):1580–1586
LaFarge A (2014) On the high line: exploring America's most original urban park. Thames and Hudson
Lee KE, Williams KJ, Sargent LD, Williams NS, Johnson KA (2015) 40-second green roof views sustain attention: the role of micro-breaks in attention restoration. J Environ Psychol 42:182–189
Lin B, Meyers J, Barnett G (2015) Understanding the potential loss and inequities of green space distribution with urban densification. Urban For Urban Green 14(4):952–958. https://doi.org/10.1016/j.ufug.2015.09.003
Maas J, van Dillen SME, Verheij RA, Groenewegen PP (2009) Social contacts as a possible mechanism behind the relation between green space and health. Health Place 15(2):586–595. https://doi.org/10.1016/j.healthplace.2008.09.006
Marselle M, Irvine K, Warber S (2013) Walking for well-being: are group walks in certain types of natural environments better for well-being than group walks in urban environments? Int J Environ Res Public Health 10(11):5603–5628
Marselle M, Irvine KN, Lorenzo-Arribas A, Warber SL (2016) Does perceived restorativeness mediate the effects of perceived biodiversity and perceived naturalness on emotional well-being following group walks in nature? J Environ Psychol 46:217–232
Mell IC (2009) Can green infrastructure promote urban sustainability? In: Proceedings of the institution of civil engineers-engineering sustainability, vol 1. Thomas Telford Ltd, pp 23–34
Miller JR (2005) Biodiversity conservation and the extinction of experience. Trends Ecol Evol 20(8):430–434
Miller JR, Hobbs RJ (2002) Conservation where people live and work. Conserv Biol 16(2):330–337
Mullaney J, Lucke T, Trueman SJ (2015) A review of benefits and challenges in growing street trees in paved urban environments. Landsc Urban Plan 134:157–166
Myers Z (2019) A healing neuropolis: nature, neuroscience and urban design. Palgrave MacMillan
Newton PW (2010) Beyond greenfield and brownfield: the challenge of regenerating Australia's greyfield suburbs. Built Environ 36(1):81–104. https://doi.org/10.2148/benv.36.1.81
Newton PW, Murray S, Wakefield R, Murphy C, Khor L, Morgan T (2011) Towards a new development model for housing regeneration in greyfield residential precincts. Australian Housing and Urban Research Institute Limited, Melbourne, Australia
Nordh H, Hartig T, Hagerhall C, Fry G (2009) Components of small urban parks that predict the possibility for restoration. Urban For Urban Green 8(4):225–235
Nouri H, Beecham S, Kazemi F, Hassanli AM (2013) A review of ET measurement techniques for estimating the water requirements of urban landscape vegetation. Urban Water J 10(4):247–259
Ossola A, Hahs AK, Livesley SJ (2015) Habitat complexity influences fine scale hydrological processes and the incidence of stormwater runoff in managed urban ecosystems. J Environ Manage 159:1–10
Palliwoda J, Kowarik I, von der Lippe M (2017) Human-biodiversity interactions in urban parks: the species level matters. Landsc Urban Plan 157:394–406. https://doi.org/10.1016/j.landurbplan.2016.09.003
Pandit R, Polyakov M, Sadler R (2014) Valuing public and private urban tree canopy cover. Aust J Agric Resour Econ 58(3):453–470
Panduro TE, Veie KL (2013) Classification and valuation of urban green spaces—a hedonic house price valuation. Landsc Urban Plan 120:119–128
Pauleit S, Ennos R, Golding Y (2005) Modeling the environmental impacts of urban land use and land cover change—a study in Merseyside, UK. Landsc Urban Plan 71(2–4):295–310
Pereira G, Christian H, Foster S, Boruff BJ, Bull F, Knuiman M, Giles-Corti B (2013) The association between neighborhood greenness and weight status: an observational study in Perth, Western Australia. Environ Health 12(1):49. https://doi.org/10.1186/1476-069X-12-49
Peschardt KK, Stigsdotter UK (2013) Associations between park characteristics and perceived restorativeness of small public urban green spaces. Landsc Urban Plan 112:26–39

Ramalho CE, Laliberté E, Poot P, Hobbs RJ (2014) Complex effects of fragmentation on remnant woodland plant communities of a rapidly urbanizing biodiversity hotspot. Ecology 95(9):2466–2478. https://doi.org/10.1890/13-1239.1

Ryu C, Kwon Y (2016) How do mega projects alter the city to be more sustainable? Spatial changes following the Seoul Cheonggyecheon restoration project in South Korea. Sustainability 8(11):1178

Schwartz MW, Jurjavcic NL, O'brien JM (2002) Conservation's disenfranchised urban poor. BioScience 52 (7):601–606

Seddon G (1994) The Australian back yard. In: Craven I (ed) Australian popular culture. Cambridge University Press, Cambridge, pp 22–35

Smith C, Clayden A, Dunnett N (2009) An exploration of the effect of housing unit density on aspects of residential landscape sustainability in England. J Urban Des 14(2):163–187. https://doi.org/10.1080/13574800802670978

Smith CA, Billig NS (2012) Public perceptions of compact suburbia in progressive, burgeoning communities. J Urban Des 17(3):313–335. https://doi.org/10.1080/13574809.2012.683401

Sugiyama T, Gunn LD, Christian H, Francis J, Foster S, Hooper P, Owen N, Giles-Corti B (2015) Quality of public open spaces and recreational walking. Am J Public Health 105(12):2490–2495

Threlfall CG, Mata L, Mackie JA, Hahs AK, Stork NE, Williams NS, Livesley SJ (2017) Increasing biodiversity in urban green spaces through simple vegetation interventions. J Appl Ecol 54(6):1874–1883

Threlfall CG, Walker K, Williams NS, Hahs AK, Mata L, Stork N, Livesley SJ (2015) The conservation value of urban green space habitats for Australian native bee communities. Biol Cons 187:240–248

Troy P (2004) Saving our cities with suburbs. In: Schultz J (ed) Griffith review: dreams of land. Griffith University, Brisbane

Tyrväinen L, Ojala A, Korpela K, Lanki T, Tsunetsugu Y, Kagawa T (2014) The influence of urban green environments on stress relief measures: a field experiment. J Environ Psychol 38:1–9

Tzoulas K, Korpela K, Venn S, Yli-Pelkonen V, Kazmierczak A, Niemela J, James P (2007) Promoting ecosystem and human health in urban areas using green infrastructure: a literature review. Landsc Urban Plan 81(3):167–178. https://doi.org/10.1016/j.landurbplan.2007.02.001

Udell T, Daley M, Johnson B, Tolley R (2014) Does density matter? The role of density in creating walkable neighbourhoods. National Heart Foundation of Australia, Melbourne

Ulrich R (1984) View through a window may influence recovery from surgery. Science 224(4647):420–421. https://doi.org/10.1126/science.6143402

van den Born RJ, Arts B, Admiraal J, Beringer A, Knights P, Molinario E, Horvat KP, Porras-Gomez C, Smrekar A, Soethe N (2018) The missing pillar: eudemonic values in the justification of nature conservation. J Environ Plan Manag 61(5–6):841–856

Voytenko Y, McCormick K, Evans J, Schliwa G (2016) Urban living labs for sustainability and low carbon cities in Europe: towards a research agenda. J Clean Prod 123:45–54

Wolch JR, Byrne J, Newell JP (2014) Urban green space, public health, and environmental justice: The challenge of making cities 'just green enough'. Landsc Urban Plan 125:234–244. https://doi.org/10.1016/j.landurbplan.2014.01.017

Wu C-D, McNeely E, Cedeño-Laurent J, Pan W-C, Adamkiewicz G, Dominici F, Lung S-CC, Su H-J, Spengler JD (2014) Linking student performance in Massachusetts elementary schools with the "greenness" of school surroundings using remote sensing. PLoS ONE 9(10):e108548

Xing Y, Jones P, Donnison I (2017) Characterisation of nature-based solutions for the built environment. Sustainability 9(1):149

Yelenik SG, Levine JM (2011) The role of plant–soil feedbacks in driving native-species recovery. Ecology 92(1):66-74. https://doi.org/10.1890/10-0465.1

Chapter 4
A Manual for Implementing Greenspace-Oriented Development

Abstract While in the previous chapter we defined Greenspace-Oriented Development (GOD) and explained the multitude of benefits it offers, here we explain how practitioners can implement GOD. We suggest a step-by-step process that aims to guide GOD implementation. These steps are: (1) select parks for upgrading; (2) upgrade parks; (3) rezone the urban precincts surrounding parks; (4) catalyze and facilitate redevelopment; (5) decentralize services infrastructure; (6) conduct needs-based assessment and equip parks, and (7) upgrade surrounding key streetscapes. This step-by-step process sets out how GOD can guide urban densification while offering multiple socio-ecological benefits through the redesign and activation of public green spaces.

Keywords Greenspace-Oriented Development · Transit-Oriented Development · Urban biodiversity · Urban consolidation · Urban design · Urban infill · Urban green space · Urban parks · Public open space

4.1 How Could GOD Help to Deliver Urban Infill in Suburban Cities?

Here, we explain how practitioners can achieve GOD in a step-by-step process (Fig. 4.1). This detailed process is likely to be particularly useful to multidisciplinary teams of practitioners involved in the planning and design of urban green spaces and residential areas. These teams may include planners, architects, landscape architects, urban designers, engineers, park managers, environmental and social science experts developers and policymakers, among others.

Alongside discussing key aspects relevant to each step, we use a hypothetical case study park for demonstration. This park is nominally 7.5 ha in size and is surrounded by suburban housing at 15 dwellings per ha and minor roads. The park is, in its "existing" state, geared towards active recreation and contains three ovals and minimal cover by mature trees and understorey plantings. Such a hypothetical park is typical of many greyfield suburbs and, therefore, our proposals for its upgrade are generalizable.

J. Bolleter and C. E. Ramalho, *Greenspace-Oriented Development*,
SpringerBriefs in Geography, https://doi.org/10.1007/978-3-030-29601-8_4

1. Select parks for upgrading
Select parks that are within a 5-minute cycle or a 20-minute walk (approximately 1,600 m) to public transport, and that are greater than 1 ha in size

2. Upgrade parks
Upgrade parks to increase their attractiveness and the socio-ecological benefits they provide

3. Rezone the parks' surrounding precincts
Rezone the 400 m urban precinct surrounding the park for higher density

4. Catalyze and facilitate redevelopment
Facilitate redevelopment catalyzed by the increase in land values resulting from both upgrading the parks and rezoning their precincts

5. Decentralize services infrastructure
Reduce the reliance of each park's higher-density urban precinct on centralized water, power, energy and wastewater infrastructure

6. Conduct needs-based assessment and equip park
Equip the "loose fit" space in relation to the requirements of the local community

7. Upgrade surrounding key streetscapes
Facilitate the upgrade of the local streets connecting GOD parks to schools, train stations or transport hubs, and main shopping areas

Fig. 4.1 Achieving GOD: This diagram sets out how practitioners can achieve GOD in a seven-step process. We explain this process in full in this chapter

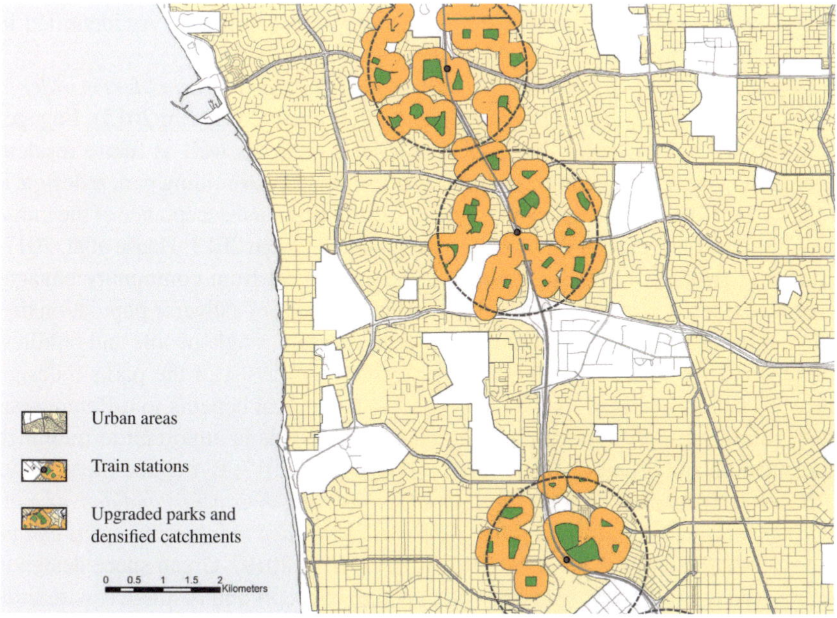

Fig. 4.2 Select parks for upgrading: In the first instance, practitioners need to identify the parks that will form the focus of GOD. We suggest that proximity to public transport and park size are two key criteria that should guide parks' prioritization

4.1.1 Step 1: Select Parks for Upgrading

In the first instance, practitioners need to identify the parks that will form the focus of GOD precincts in middle-ring suburbs. We suggest that reasonable access to public transport and park size are two key criteria that should guide parks' prioritization. We consider that reasonable access can be understood as being within a 5-minute cycle or a 15–20-minute walk (approximately 1,600 m) to train stations or rapid bus transit (Fig. 4.2). In terms of size, parks should be greater than 1 ha in area. While there is no particular reason behind this specific suggested size, larger parks have generally greater potential to provide a wider range of social (e.g. Giles-Corti et al. 2005; Sugiyama et al. 2010) and ecological (e.g. Nielsen et al. 2014) benefits than smaller parks.

4.1.2 Step 2: Upgrade Parks

In this step, practitioners redesign the selected parks to increase their attractiveness and the socio-ecological benefits they provide, which should then raise land values and encourage redevelopment of the surrounding urban precinct. As identified in Chap. 3, several key aspects should guide this step.

First, *it is crucial to engage and consult with residents and stakeholders in order to identify their preferences and needs* (Huang 2010; Shan 2012; Jim 2013). Engagement should try to identify the requirements of current, as well as future resident demographic, social and cultural groups. While this is key to guiding park redesign, it may also have a positive impact on residents' perceptions and acceptance of the transformations in their park and neighbourhood (Jim and Shan 2013; Haase et al. 2017; Smith and Billig 2012). Aside from the insight derived from community engagement processes, it is important to acknowledge the needs of a diverse population that will include older people, children, adolescents, parents, single people and families, wealthy and the poor. Recognizing this diversity will assist in the parks redesign and will help to optimize the delivery of socio-ecological benefits to different resident groups, not only for those paying the cheques. This is important to minimize gentrification and promote inclusiveness (Haase et al. 2017; Byrne and Sipe 2010).

Second, *if parks are to cater for different groups and their needs and deliver multiple benefits, then a multidisciplinary approach is needed to inform how this can be achieved* (James et al. 2009; Jim 2013; Hansen et al. 2019). Green space design is often based on traditional landscape architecture concepts and aesthetics, with little consideration of new or contrasting knowledge held by other disciplines (Hjort et al. 2018). In the redesign of GOD parks, evidence-informed knowledge from disciplines such as ecology, public health, environmental physiology and engineering is needed to establish clear relationships between the characteristics of the biophysical environment (and the changes to be brought upon it) and their functions and benefits (Hjort et al. 2018; Ko 2018). GOD proponents can achieve this, for example, by engaging and consulting with experts from different fields, including council staff, consultants and/or experts from other agencies (e.g. land management and research organizations).

Third, *park design should be a response to—and celebration of—their immediate environment* (Seddon 2005). While this is obvious, the reality is that North European aesthetic values, design and planting traditions have heavily influenced the design of Australian suburban parks. A response to the immediate *biophysical environment* in the redesign of GOD parks should entail a sensitive approach to water and fertilizer use, the preservation and enhancement of any remnant vegetation, retention and protection of mature trees (Brunner and Cozens 2013), predominant use of understorey (shrubs and herbaceous) native plants (see Webb 2013; Herd and Ivankovic-Waters 2017; Powell 2009), and enhancement of conditions that attract and sustain local biodiversity (Figs. 4.3 and 4.4). A response to the immediate *social and cultural environment* should recognize and celebrate the contemporary cultural diversity of Australian local neighbourhoods. This should entail the creation of places that celebrate Indigenous culture and foster "decolonization" and social healing (Fig. 4.5).

Fig. 4.3 Park design as a celebration of the immediate environment: The redesign of GOD parks should respond to its biophysical environment. Such may entail preserving and enhancing any remnant vegetation, retaining and protecting mature trees, and enhancing conditions that attract and sustain local biodiversity. *Source* Cristina E. Ramalho

In relation to the redesign of our hypothetical case study, below we suggest several approaches that could enhance the socio-ecological benefits provided and improve the experience of nature by residents. Readers can observe these approaches in many well-designed Australian public green spaces. We emphasize though that the suggestions presented here are merely demonstrative.[1]

Focusing on the hypothetical case study park, we propose the planting of park edges with a diverse pallet of suitable native and non-native trees, native low shrubs and herbaceous plants, so that they assume a more vegetated and diverse appearance, particularly when viewed from surrounding properties (Fig. 4.6). This planting "armature" following organic, non-rigid lines would swathe a circuitous promenade, as well as several smaller spaces designed to allow for assorted functions (Fig. 4.7). These spaces could include, for example, drainage swales for filtering and cleansing storm water flowing off the higher elevation adjacent roads, which in suitable locations would be designed to mimic natural wetlands (e.g. Guzman 2017).

[1]As mentioned above, proper consultation with the local community and stakeholders to identify their needs and perspectives, as well as a multidisciplinary approach to understand how a response to such needs can be realized in the park redesign, are fundamental to guide this step. Nonetheless, here we provide examples of what GOD proponents could achieve.

Fig. 4.4 Park design as a celebration of the immediate environment: Native plants offer a magnificent pallet of colours, forms and textures, provide ecological resources to local biodiversity (e.g. birds and insects), and require little water and fertilizer. We therefore recommend their use in plantings of GOD precincts. *Source* Cristina E. Ramalho

Individual benches could be positioned in reclusive quiet areas for rest and contemplation (Nordh et al. 2009; Nordh and Østby 2013), while in other areas clustered benches could facilitate social interaction (Grahn and Stigsdotter 2010). In both cases, benches would face away from traffic, built or busy settings, and be surrounded by native plantings, in order to promote mental restoration and connection with nature (Nordh and Østby 2013).

Practitioners could also provide areas for children's play, such as nature-based play areas, as well as picnic tables and barbecue areas (Fig. 4.8). These different areas would increase the opportunities for recreational walking, nature-based and passive recreation, and would also enhance the ecological benefits provided by the park and its ability to support biodiversity. Also, this "soft shell" of vegetation would act as a buffer between the proposed active recreation occurring in the central areas of the park and the neighbouring residents, reducing the potential for noise and sports lighting-related complaints (Lutzenhiser and Netusil 2001). We suggest that the park could have a low, visually unobtrusive fence, to prevent children and dogs from

Fig. 4.5 Park design as a celebration of the immediate environment: The redesign of GOD parks should also respond to the social and cultural environment. The parkland of the Optus Stadium in the Burswood Peninsula, Perth, offers multiple examples of how park design can celebrate Indigenous culture. Indigenous elements are embedded in the artwork (such as the message stick in the foreground), built infrastructure, plant pallets, local materials and education signage. *Source* Cristina E. Ramalho

running on to the surrounding roads, and to direct entry to the park at specific points. Entry points could be located at street intersections to help deviate park foot traffic away from fringing residential buildings.

While the heart of the park is to remain open, we propose that practitioners consolidate the three sports ovals into one single oval, which would be re-turfed with a hybrid species that allows for greater frequency of sporting and community uses (e.g. festivals and markets). The consolidation of the oval area is not meant to detract from the important social and recreational functions of team sports but to provide a greater number of passive (and active) recreational pursuits. Also, ovals located in other urban parks that are not in zones of proposed GOD densification could specifically cater to major active sports events.

The area freed up by the consolidated ovals would be a flexible "loose-fit" space (Franck and Stevens 2007; Thompson 2002) allowing for a variety of functions developed in relation to shifting community preferences. Researchers have suggested that "loose spaces offer a freedom of choice of activities and more means of carrying

Fig. 4.6 Park upgrades, before and after: We propose the planting of park edges with a diverse pallet of suitable native and non-native trees, low native shrubs and herbaceous plants

Fig. 4.7 Park upgrades, before and after: The planting "armature" framing the park would swathe a circuitous promenade and several smaller spaces designed to allow for assorted functions. Such could include drainage swales and wetlands (as shown) for filtering and cleansing storm water and providing habitat for biodiversity. Images by Robert Cameron

Fig. 4.8 Park upgrades, before and after: Practitioners could provide areas for children's play, such as nature-based play areas, as well as picnic tables and barbecue areas. These different areas would increase the opportunities for recreational walking, passive recreation and social interaction. Images by Robert Cameron

them out", and that such spaces are open to appropriation by the local community (Franck and Stevens 2007; Thompson 2002).[2] The exact use of this space could be established after significant residential densification has occurred (step 4) through a comprehensive needs-based assessment.

4.1.3 Step 3: Rezone the Urban Precinct Surrounding Parks

In this step, planners should rezone the 400 m urban precinct surrounding the park. This precinct is commensurate with the area in which the park's upgrade is likely to lift property values (Crompton 2005; Lutzenhiser and Netusil 2001). We also intend for this to be the area to undergo significant infill development.

For the hypothetical case study park, we visualize the rezoning of the surrounding urban precinct into three zones of differing density: 40 dwellings per ha furthest from the park, 60 dwellings per ha mid-way, and 80 dwellings per ha closest to the park. These zoning densities correlate to semi-detached dwellings, row houses, and low-rise apartments, respectively (Kellet and Rofe 2009). If substantially achieved, these zoning densities would increase the total population of the study area from 1,500 to around 6,500 people.[3] To give an idea of the capacity of GOD, Perth (for instance) has 420 suitable greyfield parks; if these parks and their surrounding urban precincts were developed using GOD, it could yield well over 2 million new infill dwellings. This is substantially more than the state government's infill dwelling target of 121,000 dwellings for the same area by 2031 (Western Australian Department of Planning 2010).

We also suggest that practitioners should rezone the areas immediately adjacent to the park to allow retail, food and beverage, and office[4] land uses, as well as residential. The zoning of the walkable precinct should also allow other community services and functions, such as retirement homes and childcare centres, which have potential synergies with upgraded green spaces. This will ensure that GOD parks are, as Jane Jacobs evoked, "where life swirls – where there is work, cultural, residential and commercial activity – as much as possible of everything that cities can offer" (Jacobs 1962).

[2]This is particularly true for children. While the literature and guidelines often focus on more structured features of parks and open space (such as playground equipment and paths), from a broader child development perspective, providing opportunities for creative play, exploring, make-believe, contact with different textures and exposure to nature are all critical, and often missing in the more typical "plastic fantastic" playgrounds of today (Wood 2008).

[3]This presumes that the precinct was 75% redeveloped at the zoned densities and that each dwelling contained a household of two people.

[4]A disincentive to working from home is a lack of social interaction. Co-working might offer a middle ground where people could share and rent a desk space at a co-working office close to their home instead of going to the "city" (Gladstone et al. 2016).

Again, we emphasize that the suggestions we have presented here in Step 3 are merely demonstrative and would depend on community sentiment towards infill development, land values and existing lot sizes, amongst other factors.

4.1.4 Step 4: Catalyze and Facilitate Redevelopment

We suggest that the combined effect of both upgrading the parks and rezoning their surrounding urban precincts is likely to catalyze the redevelopment of the area, due to an increase in adjacent private land values. Studies using hedonic valuation techniques, which estimate the influence of the locality and house attributes on housing prices, have consistently indicated that high-quality parks and lakes raise property values in adjacent areas, but that sports fields do not have the same effect (Panduro and Veie 2013; Brander and Koetse 2011; Crompton 2005; Lutzenhiser and Netusil 2001).

In Portland, US, Lutzenhiser and Netusil (2001) found that parks with more than 50% of their area covered with natural vegetation, and used mostly for biodiversity conservation and nature-oriented recreation (e.g. walking, wildlife viewing), had the largest positive effect on the sale price of houses located within a 450 m radius. In contrast, parks with more than 50% of their area manicured or landscaped, and used primarily for active recreation (e.g. ball fields and sports courts), had the lowest positive effect on property values surrounding the open space. Very similar results were observed by Crompton (2005) in a two-decade review of the impact of parks on property values in the US. In Perth, Pandit et al. (2014) found that bush reserves, lakes and golf courses had a positive impact on property prices, but the same was not observed for sports reserves. The authors further noted that, on average, the property price premium increased by AU$14,500 for a 10% increase in tree canopy cover on adjacent public space.

Given the above data, upgrading suburban parks using a GOD approach should raise nearby real estate values. This will provide local governments with greater resources for park maintenance[5] and should stimulate redevelopment (Pracsys 2012; Newton et al. 2011), which in combination with increased residential zoning densities should deliver greater urban densification. The stimulation of redevelopment is, in part, explained by the Index of Property Redevelopment Potential. This index indicates that by increasing the value of a parcel of land in relation to the value of the house, redevelopment is encouraged (Newton et al. 2011). This is crucial because if development is not feasible, quite simply, nothing will happen.

As redevelopment of the park's precinct occurs, it is crucial that controls are in place to ensure that background infill, as we discussed in Chap. 2, does not run riot. We suggest that practitioners could use the following key controls to support the redevelopment of the park's urban precinct. We have developed these controls based

[5]Increases in property value increase the revenue to local governments from property rates (Pauli and Boruff 2016).

on interviews with property developers, architects, community representatives, real estate experts and planners.

4.1.4.1 Mandate Minimum Lot Sizes

Practitioners should constrain background, low-density infill by establishing minimum lot sizes—and densities—for redevelopment in the GOD precinct. While it will vary from area to area, a possible minimum lot size of around 1,200 m^2 will essentially preclude any infill unless developers can amalgamate adjoining lots—the typical lot being 1,000 m^2 or less in greyfield suburbs (Fig. 4.9). Short of mandating

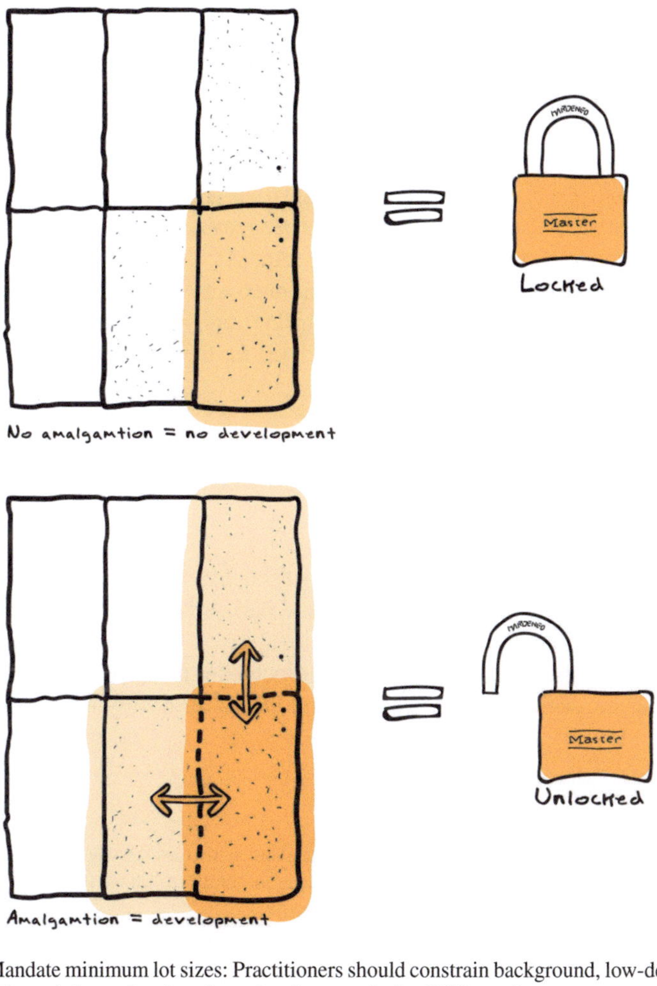

Fig. 4.9 Mandate minimum lot sizes: Practitioners should constrain background, low-density infill by establishing minimum lot sizes for redevelopment in the GOD precincts

minimum lot sizes, local governments could offer greater incentives to encourage land amalgamation, such as zonings that allow higher densities for amalgamated lots. The importance of enforcing (or incentivizing) lot amalgamation in the park precinct is that it creates larger development sites. This in turn provides building designers with the additional room needed to respond to solar orientation and existing trees, and achieve higher buildings while reducing the perceived density.[6]

4.1.4.2 Protect and Densify the Urban Forest

As explained in Chap. 2, one of the reasons local communities may oppose urban infill is because they perceive it to be an assault on the "leafy greenness" of their neighbourhoods. A control that could partially respond to this, while ensuring the presence of trees in the park's urban precinct, would be one that retains and protects mature trees, and establishes minimum areas for deep soil zones to allow for the healthy growth of new trees.

One way of aiding tree retention is to allow buildings to be built right up to lot boundaries on several sides to produce more consolidated internal "courtyard" areas of private open space for tree planting (Figs. 4.10 and 4.11). This would combat the narrow "corridors" of private open space between a lot boundary and a building commonly produced by regulated building setbacks, for example, in the Western Australian Residential Design Codes.[7] Practitioners should also support the delivery of more slender buildings with smaller footprints, and place outdoor living spaces on roof terraces, thus freeing up the ground level for tree and understorey planting[8] (Figs. 4.12 and 4.13).

4.1.4.3 Develop Using Environmental and Social Sustainability Approaches

The design of the higher-density urban precinct should use green building technology and embed greening solutions that soften the park–urban transition, create visual connectivity and quality, and provide additional green space benefits to residents. The design should also actively mitigate the potential negative impacts of higher-density housing, such as crowding, noise and indoor air quality issues (Giles-Corti

[6]The effect of maintaining mature trees is to decrease the "perceived density" of development (as opposed to the actual density), which should reduce community resistance (Cheng and Steemers 2010).

[7]Small trees typically require an area of 3.5 × 3.5 m (Department of Planning 2016). As such, these narrow corridors effectively preclude tree planting (large or small).

[8]Practitioners should carefully choose plants that do not drop limbs, do not have large and shallow rooting systems, and have low flammability if in bushfire-prone areas.

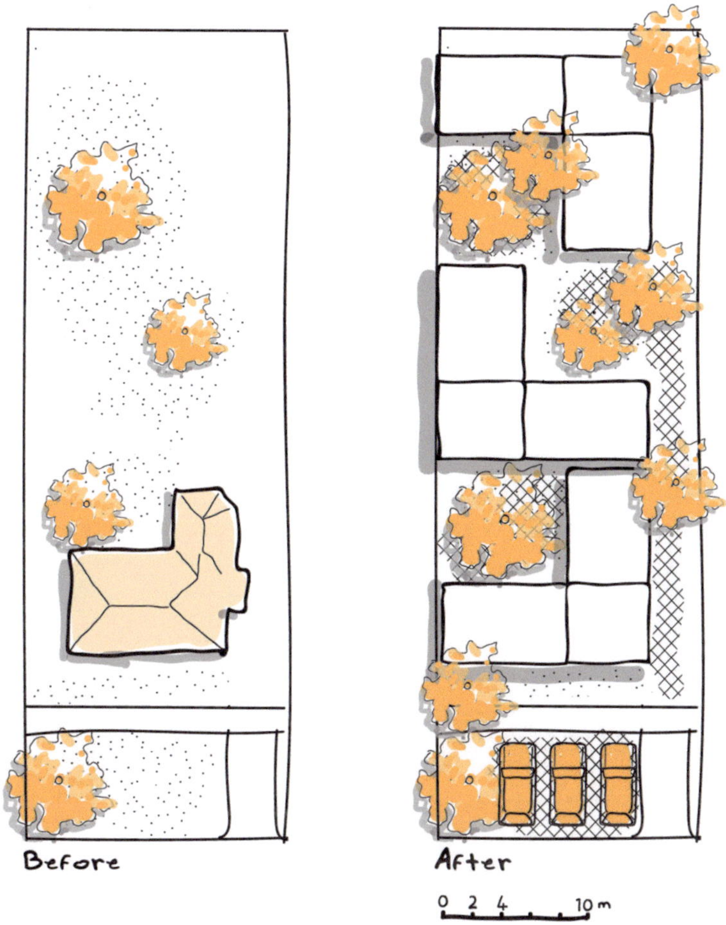

Fig. 4.10 Protect and densify the urban forest: One way of aiding retention of mature trees is to allow buildings to be built right up to lot boundaries on a number of sides to produce more consolidated internal "courtyard" areas of private open space for tree planting

et al. 2012). Equally important, developers should deliver a range of housing sizes, styles and price options, so that the higher-density urban precinct can cater for all ages and stages of life, as well as income levels. This would minimize gentrification and promote social inclusiveness.

4.1.5 Step 5: Decentralize Services Infrastructure

In this step, practitioners should reduce the reliance of each park's higher-density urban precinct on centralized water, power, energy and wastewater management

Fig. 4.11 Protect and densify the urban forest: Practitioners should design courtyard dwellings that are arranged around existing mature trees (and deep soil zones) and frame views of vegetation

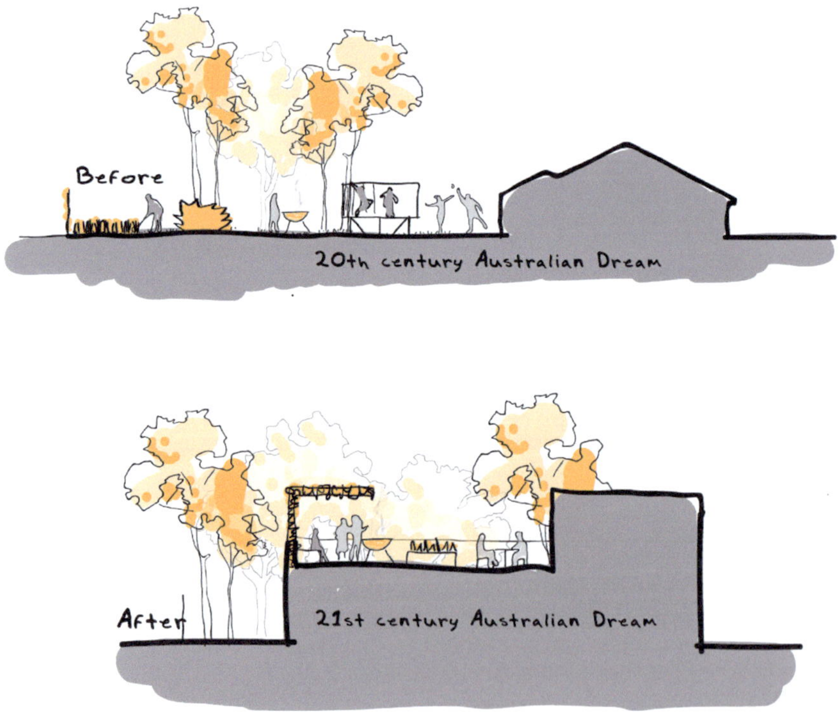

Fig. 4.12 Protect and densify the urban forest: Practitioners should also support the delivery of more slender buildings with smaller footprints, and place outdoor living spaces on roof terraces, thus freeing up the ground level for tree and understorey planting

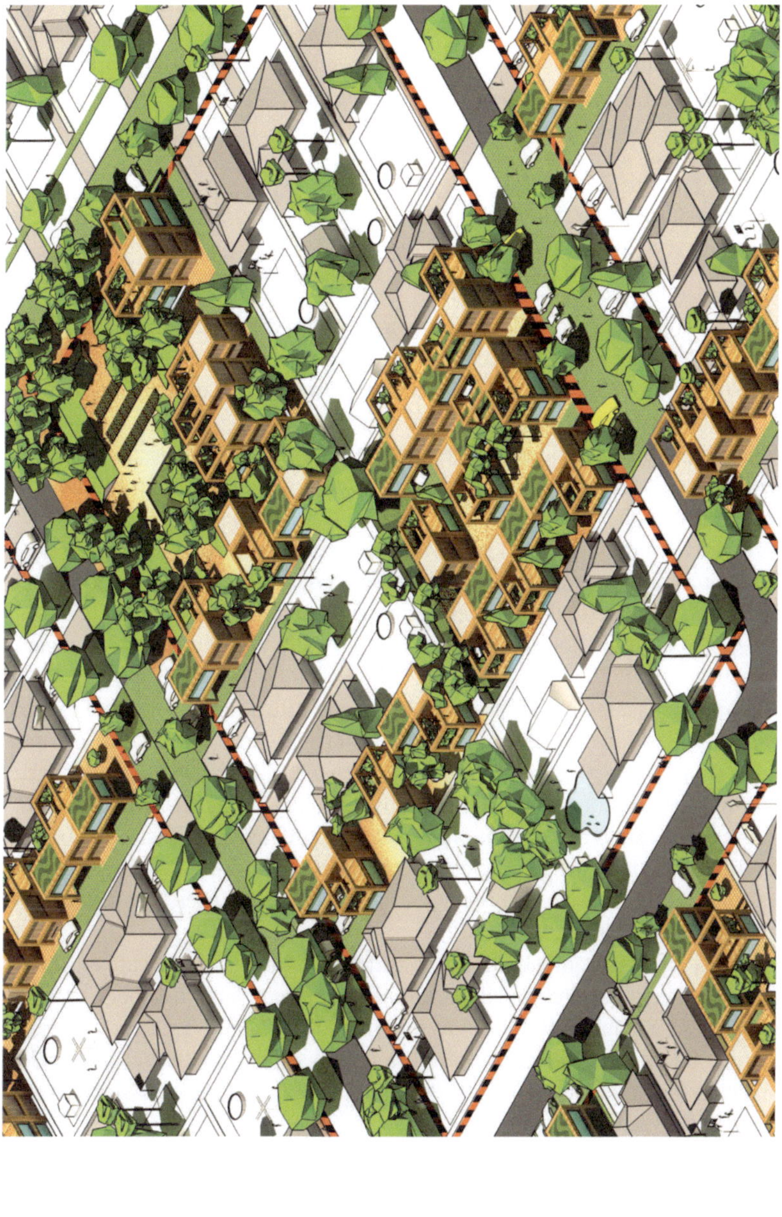

Fig. 4.13 Protect and densify the urban forest: With appropriate building types and appropriate tree species (e.g., relatively small trees without large shallow rooting systems), practitioners can increase the density of urban form and the urban forest simultaneously

infrastructure.[9] Decentralized infrastructure, in this context, could include wastewater treatment facilities to clean and recycle grey and black water from the higher-density area. Tanks below or above ground could store such water for irrigation in the park, urban precinct and surrounding streetscapes. Facilities for green waste collection and composting could also be made available (Grace 2013). In this respect, the upgraded park and its densified urban precinct would function as a cell of decentralized infrastructure, which is to some extent free from the inefficiencies of typically ageing, centralized infrastructure (Grace 2013; Newman et al. 2009).

4.1.6 Step 6: Conduct Needs-Based Assessment and Equip Park

In this step, after significant densification has occurred in the park's urban precinct, we suggest local governments and/or community groups equip the "loose-fit" space in the park so that it provides additional recreational amenity to the local community (Fig. 4.14). At this point, practitioners should conduct a needs-based assessment to establish the recreational facilities and equipment required to activate the park. A needs-based assessment is important because, as Byrne and Sipe (2010) explained, "there is no typical higher-density resident". Indeed, higher-density residents vary in age, income, race/ethnicity, household composition, family status and the like. Practitioners could partly conduct this needs-based assessment using a smartphone application and/or website that makes engagement easier, especially if it includes three-dimensional visualization tools and online consultation (Kelly and Donegan 2015).

The needs-based assessment should lead to the identification of a diverse range of activities and uses for the loose-fit space (Kellet and Rofe 2009). These uses could include skate-able spaces, informal BMX areas, basketball and netball rings, soccer goals and spaces (not a whole field), innovative play areas, fitness equipment, enclosed dog exercise areas and community vegetable gardens (Suter Planners 2011). Complementing these uses are the typically organized team sports that the retained oval caters for, and the passive recreation and nature-oriented uses that the park's "armature" redesign caters for.

At this stage, local governments should consider including a private café or kiosk to further activate the place and provide a revenue stream. A large number of parks in Europe and the US contain food stands, kiosks, cafés, restaurants, beer gardens, equipment hire facilities and other applicable commercial uses that can provide revenue to local governments for ongoing park maintenance (Byrne and Sipe 2010).

[9]Such parks will need to be larger than 2 ha. Therefore, not all parks will be suitable for the decentralization of infrastructure.

Fig. 4.14 Conduct needs-based assessment and equip park: After significant densification has occurred in the park's precinct, we suggest local governments and/or community groups equip the "loose-fit" space in the park (pictured) so that it provides additional recreational amenity to the local community

4.1.7 Step 7: Upgrade Surrounding Key Streetscapes

In this last step, practitioners should oversee the upgrade of the local streets connecting GOD parks to schools, train stations or transport hubs, and main shopping areas (Figs. 4.15, 4.16 and 4.17). These streets should be conceptualized as shared zones promoting active transport (walking and cycling), as well as emerging transport types, such as neighbourhood electric vehicles, mobility scooters (gophers), e-bikes and e-scooters (Atkins 2016), while reducing the speed and impact of cars.[10] Connecting streetscapes should also provide shared community facilities, such as small playgrounds, community gardens, benches and other designed street furniture, as well as appropriate canopy cover (Coutts and Tapper 2017; Sanusi et al. 2017) and understorey plantings. If space is tight on the ground, then green spaces can be encapsulated within the built environment itself, namely in building facades and walls.

Practitioners can facilitate the upgrading of these streets to be shared zones by reconfiguring the roadway so that it "meanders", creating usable areas of public green spaces either side, rather than merely bisecting the road reserve directly down

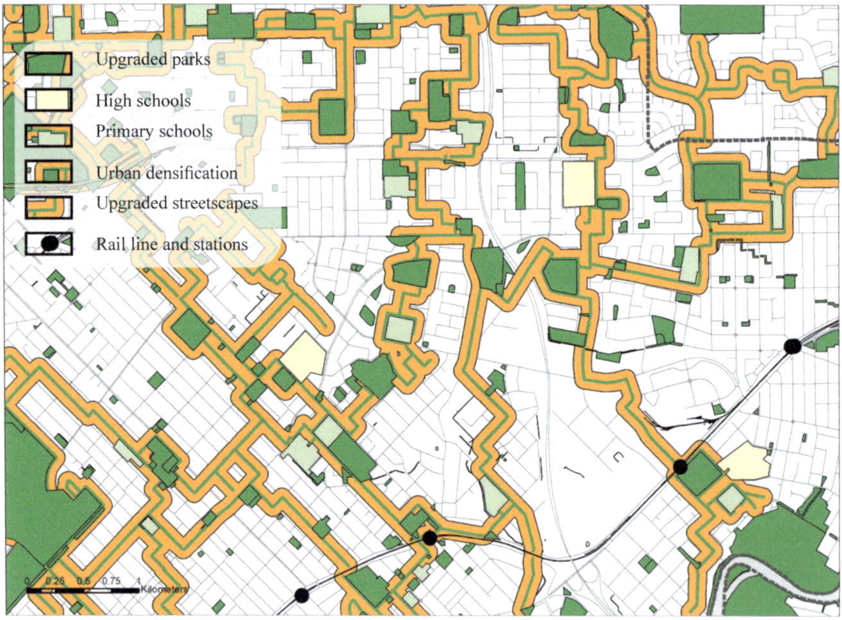

Fig. 4.15 Upgrade surrounding key streetscapes: In this last step, practitioners should oversee the upgrade of the local streets (shown in orange) connecting the upgraded parks to schools, train stations or transport hubs, and main shopping areas

[10]This will be helped by the widespread adoption of Fully Automated Vehicles, which will be generally safer than human drivers (Gladstone et al. 2016).

Fig. 4.16 Upgrade surrounding key streetscapes, before and after: These upgraded streets should be conceptualized as shared zones promoting active transport, as well as emerging transport types, such as e-bikes and e-scooters, while reducing the speed and impact of cars

Fig. 4.17 Upgrade surrounding key streetscapes: Connecting streetscapes should provide shared community facilities, as well as appropriate canopy cover

the middle. This proposition is nothing new. The shared street (Woonerf) concept has been adopted in guidelines and regulations originally in the Netherlands (1976) and also in many other countries: Germany, England, Sweden and Denmark, France, Japan, Israel and Switzerland (Ben-Joseph 2007).

Planners should complement the increased open space amenity along connecting streetscapes with an increase in residential density. Again, this redevelopment will be incentivised because greater canopy and green cover will increase adjacent property values. In a study set in Perth, Pandit et al. (2013) found that street trees adjacent to a house produced "positive and sizable effects" on the house's sale price.[11] As such, further street tree planting, and by extension streetscape upgrades as suggested here, would lift property values, increase development feasibility and give residents a "hip-pocket" reason to support their redevelopment.

While good connections to public transport remain important, cars are unavoidable for the foreseeable future (Hagan 2017). As such, practitioners need to design some road sections within GOD precincts with perpendicular parking to efficiently contain

[11] They concluded that the marginal implicit price of a broad-leaved tree on the street verge was about AU$17,000, which corresponds to a just over 4% increase in the median value of the property.

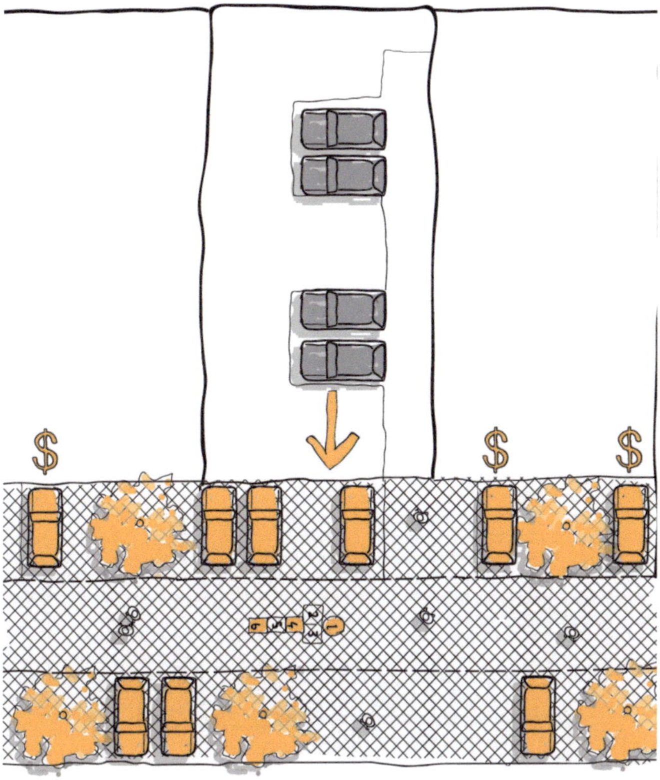

Fig. 4.18 Upgrade surrounding key streetscapes: While good connections to transit remain important, cars are unavoidable for the near future. As such, planning teams need to design some road sections within GOD precincts with perpendicular parking to efficiently contain parked cars

the parking required for residents living at higher densities. This avoids the waste of area involved in including substantial car parking on private lots (Figs. 4.18, 4.19 and 4.20). Over time, an autonomous Central Area Transit bus connecting GOD precincts with major public transport hubs, and technological developments including ride-sharing apps and car-sharing arrangements could reduce the need for this car parking.

4.2 Conclusion

This chapter has set out a process by which multidisciplinary teams can deliver GOD in Australian suburban cities. The process remains hypothetical as planners and developers have not yet delivered any GOD specifically under our model. We intend in time, however, to document the urban projects which follow this process to

Fig. 4.19 Upgrade surrounding key streetscapes: Upgrades could include efficient right-angle parking, multi-functional shared vehicular and pedestrian spaces and an increased number of street trees

better understand how practitioners can tweak the process for improved outcomes. The next chapter summarizes the key points of the book, and sets out the implications of continuing with our current approaches to urban infill, considering Australia's rapidly increasing population. It also reflects on what the GOD alternative can offer as part of this broader perspective.

Fig. 4.20 Upgrade surrounding key streetscapes: Efficient right-angle parking on verges can reduce the substantial area required for cars when car parking is provided on private lots. This allows practitioners and developers to deliver efficient courtyard dwellings and maintain mature trees

References

Atkins M (2016) Boomers in boomtown: age-friendly planning in Australia. In: Biermann S, Olaru D, Paul V (eds) Planning boomtown and beyond. UWA Publishing, Perth, pp 70–101

Ben-Joseph E (2007) Changing the residential street scene: adapting the shared street (Woonerf) concept to the suburban environment. J Am Plan Assoc 61(4):504–515

Brander LM, Koetse MJ (2011) The value of urban open space: meta-analyses of contingent valuation and hedonic pricing results. J Environ Manage 92(10):2763–2773. https://doi.org/10.1016/j.jenvman.2011.06.019

Brunner J, Cozens P (2013) 'Where have all the trees gone?' Urban consolidation and the demise of urban vegetation: a case study from Western Australia. Plan Pract Res 28(2):231–255. https://doi.org/10.1080/02697459.2012.733525

Byrne J, Sipe N (2010) Green and open space planning for urban consolidation—a review of the literature and best practice. In: Urban research program, vol issues paper 11. Griffith University, Brisbane, Australia

Cheng V, Steemers K (2010) Perception of urban density. In: Mostafavi M, Doherty G (eds) Ecological urbanism. Lars Muller Publishers, Baden, pp 476–481

Coutts A, Tapper N (2017) Trees for a cool city: guidelines for optimised tree placement. Cooperative Research Centre for Water Sensitive Cities, Melbourne, Australia

Crompton JL (2005) The impact of parks on property values: empirical evidence from the past two decades in the United States. Manag Leis 10(4):203–218

Department of Planning (2016) Apartment design. Department of Planning, Department of Finance, Office of the Government Architect, Western Australian Planning Commission, Perth

Franck K, Stevens Q (eds) (2007) Loose space: possibility and diversity in urban life. Routledge, London

Giles-Corti B, Broomhall MH, Knuiman M, Collins C, Douglas K, Ng K, Lange A, Donovan RJ (2005) Increasing walking: how important is distance to, attractiveness, and size of public open space? Am J Prev Med 28(2):169–176

Giles-Corti B, Ryan K, Foster S (2012) Increasing density in Australia: maximising the health benefits and minimising the harm. National Heart Foundation of Australia, Melbourne

Gladstone L, Sun Y, Taplin J (2016) Hype or hope—can future transport technologies ease congestion? In: Biermann S, Olaru D, Paul V (eds) Planning boomtown and beyond. UWA Publishing, Perth, pp 574–591

Grace B (2013) Interview. Perth

Grahn P, Stigsdotter UK (2010) The relation between perceived sensory dimensions of urban green space and stress restoration. Landsc Urban Plan 94(3):264–275

Guzman CB (2017) Suburban wetlandia. In: Berger A, Kotkin J, Guzman C (eds) Infinite suburbia. MIT, Boston, pp 478–495

Haase D, Kabisch S, Haase A, Andersson E, Banzhaf E, Baró F, Brenck M, Fischer LK, Frantzeskaki N, Kabisch N, Krellenberg K, Kremer P, Kronenberg J, Larondelle N, Mathey J, Pauleit S, Ring I, Rink D, Schwarz N, Wolff M (2017) Greening cities—to be socially inclusive? About the alleged paradox of society and ecology in cities. Habitat Int 64:41–48. https://doi.org/10.1016/j.habitatint.2017.04.005

Hagan S (2017) Metabolic suburbs or the virtue of low densities. In: Berger A, Kotkin J, Guzman C (eds) Infinite suburbia. MIT, Boston, pp 468–477

Hansen R, Olafsson AS, van der Jagt AP, Rall E, Pauleit S (2019) Planning multifunctional green infrastructure for compact cities: what is the state of practice? Ecol Ind 96:99–110

Herd K, Ivankovic-Waters J (2017) Native: art and design with Australian plants. Thames and Hudson, Australia

Hjort M, Martin WM, Stewart T, Troelsen J (2018) Design of urban public spaces: intent vs. reality. Int J Environ Res Public Health 15(4):816

Huang S-CL (2010) The impact of public participation on the effectiveness of, and users' attachment to, urban neighbourhood parks. Landsc Res 35(5):551–562

Jacobs J (1962) The death and life of great American cities. Jonathon Cape, London
James P, Tzoulas K, Adams M, Barber A, Box J, Breuste J, Elmqvist T, Frith M, Gordon C, Greening K (2009) Towards an integrated understanding of green space in the European built environment. Urban For Urban Green 8(2):65–75
Jim CY (2013) Sustainable urban greening strategies for compact cities in developing and developed economies. Urban Ecosyst 16(4):741–761. https://doi.org/10.1007/s11252-012-0268-x
Jim CY, Shan X (2013) Socioeconomic effect on perception of urban green spaces in Guangzhou, China. Cities 31:123–131. https://doi.org/10.1016/j.cities.2012.06.017
Kellet J, Rofe M (2009) Creating active communities: how can open and public spaces in urban and suburban environments support active living? Institute for Sustainable Systems and Technologies, University of South Australia, Adelaide
Kelly J-F, Donegan P (2015) City limits: why Australian cities are broken and how we can fix them. Melbourne University Press, Melbourne
Ko Y (2018) Trees and vegetation for residential energy conservation: a critical review for evidence-based urban greening in North America. Urban For Urban Green 34:318–335
Lutzenhiser M, Netusil NR (2001) The effect of open spaces on a home's sale price. Contemp Econ Policy 19(3):291–298
Newman P, Beatley T, Boyer H (2009) Resilient cities. Island Press, Washington
Newton P, Murray S, Wakefield R, Murphy C, Khor L-A, Morgan T (2011) Towards a new development model for housing regeneration in greyfield residential precincts. Australian Housing and Urban Research Institute, Swinburne/Monash
Nielsen AB, Van Den Bosch M, Maruthaveeran S, van den Bosch CK (2014) Species richness in urban parks and its drivers: a review of empirical evidence. Urban Ecosyst 17(1):305–327
Nordh H, Hartig T, Hagerhall C, Fry G (2009) Components of small urban parks that predict the possibility for restoration. Urban For Urban Green 8(4):225–235
Nordh H, Østby K (2013) Pocket parks for people—a study of park design and use. Urban For Urban Green 12(1):12–17
Pandit R, Polyakov M, Sadler R (2014) Valuing public and private urban tree canopy cover. Aust J Agric Resour Econ 58(3):453–470
Pandit R, Polyakov M, Tapsuwam S, Moran T (2013) The effect of street trees on property value in Perth, Western Australia. Landsc Urban Plan 110:134–142
Panduro TE, Veie KL (2013) Classification and valuation of urban green spaces—a hedonic house price valuation. Landsc Urban Plan 120:119–128
Pauli N, Boruff B (2016) Natural environments, ecosystem services and green infrastructure: planning for perth's 'green' matrix. In: Biermann S, Olaru D, Paul V (eds) Planning boomtown and beyond. UWA Publishing, Perth, pp 238–276
Powell R (2009) Leaf and branch—trees and tall shrubs of Perth. Department of Environment and Conservation, Perth, Australia
Pracsys (2012) Fremantle traffic bridge; economic and social benefit assessment. Pracsys, Perth
Sanusi R, Johnstone D, May P, Livesley SJ (2017) Microclimate benefits that different street tree species provide to sidewalk pedestrians relate to differences in Plant Area Index. Landsc Urban Plan 157:502–511
Seddon G (2005) The old country—Australian landscapes, plants and people. Cambridge Press, Melbourne, Australia
Shan X-Z (2012) Attitude and willingness toward participation in decision-making of urban green spaces in China. Urban For Urban Green 11(2):211–217
Smith CA, Billig NS (2012) Public perceptions of compact suburbia in progressive, burgeoning communities. J Urban Des 17(3):313–335. https://doi.org/10.1080/13574809.2012.683401
Sugiyama T, Francis J, Middleton NJ, Owen N, Giles-Corti B (2010) Associations between recreational walking and attractiveness, size, and proximity of neighborhood open spaces. Am J Public Health 100(9):1752–1757
Planners Suter (2011) Principles and guidelines: best practice open space in higher density developments project. City of Charles Sturt, Adelaide

Thompson CW (2002) Urban open space in the 21st century. Landsc Urban Plan 60(2):59–72. https://doi.org/10.1016/S0169-2046(02)00059-2
Webb M (2013) Australian native plants: the Kings Park experience. CSIRO Publishing, Perth, Australia
Western Australian Department of Planning (2010) Directions 2031 and beyond: metropolitan planning beyond the horizon. Department of Planning, Perth
Wood L (2008) Action for young Australians report: parks and open space for health and wellbeing of children and young people. The Australian Research Alliance for Children and Youth, Canberra

Chapter 5
Conclusion

Abstract This chapter briefly summarizes the key points of the book. We then direct researchers to potentially fertile areas for future research. These include a systematic evaluation of community sentiment in relation to urban infill strategies, and the potential effects of emerging transport types on Transit-Oriented Development, amongst others. We then speculate on the possible societal and environmental implications of both our current strategies for urban densification and the urban sprawl that they inadvertently precipitate. We discuss this in the context of cities in Australia, and globally, many of which demographers project to grow significantly in this twenty-first century.

Keywords Greenspace-Oriented Development · Transit-Oriented Development · Urban consolidation · Urban infill · Urban sprawl · Suburban expansion · Population growth

5.1 Summary of Key Points

In this book, we examined the shortfalls of exclusively using Transit-Oriented Development (TOD) to support urban densification, with a focus on Australian cities. We then proposed a complementary approach, Greenspace-Oriented Development (GOD).

In Chap. 2 of this book, we reviewed TOD and Activity Centre planning in Australian cities. While the principles of TOD are well established, their application in Australia, namely, in Perth, Adelaide, Melbourne, Sydney and South East Queensland, has delivered comparatively little urban density co-located with public transport (Kelly and Donegan 2015; Goodman and Moloney 2004; Burton 2017; Goodman 2017; Randolph et al. 2017). This is even though such policies have been in place for a considerable amount of time, in some cases since the 1980s. Indeed, Australian cities continue to have some of the lowest residential densities in the world (Hurley et al. 2017).

Much of the infill development in Australian cities is still comparatively low-density, dispersed "background infill", not near effective public transport. This form of infill can have detrimental effects on urban forests, street interfaces and private

J. Bolleter and C. E. Ramalho, *Greenspace-Oriented Development*,
SpringerBriefs in Geography, https://doi.org/10.1007/978-3-030-29601-8_5

open space provision, all of which can negatively affect local communities and has led to a public sullenness around infill development (Kelly and Donegan 2015). These poor outcomes and the resulting community backlash subsequently curtail the delivery of further infill projects. Chapter 2 concluded that we have yet to accept the limitations of TOD strategies and learn from both its successes and failures. However, readers should be mindful that achieving urban densification in suburban cities is a difficult thing. We do not mean to denigrate the sincere efforts of practitioners to achieve TOD; rather, we seek to point out the need for complementary alternatives to TOD to improve the likelihood of curtailing sprawl.

In Chap. 3, we set out a complementary vision to achieve urban densification and reduce urban sprawl in Australian cities, and we presented the theoretical underpinnings of GOD. At its foundation, a GOD approach builds upon the now well-recognized importance of urban green spaces in delivering an array of benefits to urban dwellers, and most importantly, in underpinning approaches for greater sustainability and livability in cities. We discussed how GOD weaves together the positive aspects of suburbia (access to open space and nature) and those of good quality medium-density urban infill (public transport accessibility, housing affordability, good urban design, and improved public and cultural amenity).

In Chap. 4, we showed how practitioners can put that vision into practice, providing a step-by-step manual for implementing GOD. This direct instruction is important because greening solutions and urban densification are fundamental to deal with the challenges of unprecedented urban and population growth, as well as climate change.

5.2 Future Research

There is still much for researchers to do in the areas we have "touched on" in this book. An initial research project should test our GOD principles—and those of TOD—in relation to community sentiment in key areas of Australian cities. In grappling with the thorny issue of community resistance, planners should be careful to listen to the "real people" of Australia's cities to understand their "needs, wants, capabilities and fears", rather than falling into a trap of unquestioningly applying an apparent cure-all planning ideology. This has (to some degree) occurred with TOD—an approach which might have convinced planners but hasn't necessarily convinced communities. As Patrick Troy remind us, pursuit of "architectural and lifestyle fashions imported from other cultures that are not grounded in the lived experience or aspirations of the majority of Australians" only waste a lot of the planners' time and government money (Troy 2004). The enduring popularity of the "suburban dream" in Australia and entrenched community resistance to urban infill attests to this. Bruegmann (2017) reinforces the importance of working with, rather than against the prevailing sensibilities:

> To make a real contribution to the emerging urban pattern, it would probably help for architects, planners and public policymakers to move away from their fixation on the forms of the past, traditional aesthetic notions, and attempts to build cities to accommodate existing technology and ways of life.

Cities are human systems first and built environments second. Too often planning experts use the "built environment" rubric inappropriately to direct discussion and urban planning (Gleeson et al. 2010). As Brendan Gleeson et al. reminds us, "contemplation of the built environment is surely critical but should flow from, and not precede, this appreciation" (Gleeson et al. 2010). We believe, and the data supports, that GOD could be an appropriate, yet adaptable expression of this democratic wellspring; however, planners and policymakers require in-depth research to confirm this.

Second, TOD planning should be subject to a further fine-grained analysis. According to the various State and Territory policies across Australia there are 343 Activity Centres identified for infill development nationwide (Bolleter and Weller 2013). While in Chap. 2 we briefly set out some of the problems these Activity Centres have faced in implementation, what planners need is a national scale review to see whether planning is achieving the policy objectives for Activity Centres.

Thirdly, emerging trends also threaten TOD planning and are worthy of further research. As we discussed briefly, the predicted widespread adoption of Fully Automated Vehicles (FAVs) will mean that co-locating urban densification with public transport is less of a concern. Moreover, the World Health Organization's projections for epidemics to "commonly occur" in the twenty-first century may mean that high-density living above and adjacent to crowded train stations is less than ideal from a population health perspective. Researchers should explore, through scenario planning, what potential impacts these drivers could have for TOD Activity Centre development.

Fourthly, one of the major blockages to transformational urban change has been a lack of design vision that can capture the public imagination for more sustainable urban futures (Dovey and Woodcock 2014). While in this book we have visualized the potential of GOD to engage residents of Australia's capital cities for an alternative urban future, there is more work for researchers and designers to do in this space. Indeed, there is a lacuna of visionary design and planning work in Australia which means that urban planning (often) becomes driven by ideology rather than imagination (Dovey and Woodcock 2014).

Finally, GOD proponents need to get together and build some GODs so that researchers like ourselves can critically pick through them to discern successes and failings. Moreover, in this book we have scoped the potential of GOD specifically in greyfield, middle-ring suburbs. Researchers could direct future research to considering how planners could apply GOD principles in both inner city and greenfield settings.

5.3 Implications

As we set out in the introductory chapter, all Australian State and Territory capital cities have planning policies and targets to achieve urban densification. The achievement of this planning objective is crucial to managing Australia's rapidly growing population.[1] At the same time, Australian cities are generally not meeting their infill targets, particularly as they relate to Activity Centres (Bolleter and Weller 2013), which is perpetuating problematic sprawling city forms. If Australian cities continue not to meet these targets, their rapid population growth will mean they sprawl in what experts recognize as a typically unhealthy, costly, unsustainable and unproductive manner. This will become a particular issue as Sydney and Melbourne develop into mega-cities with populations of 10 million or more—with the attendant livability issues such as congestion, lack of housing affordability, pollution and declining access to nature—amongst others (Bolleter and Weller 2013) (Fig. 5.1).

Given this projected population growth, our capacity to deliver high amenity, resilient and connected urban densification will be one challenge that will define the

Fig. 5.1 The specter of sprawl: If Australian cities do not meet their infill targets, their rapid population growth will mean they continue to sprawl in what experts recognize as a typically unhealthy, costly, unsustainable and unproductive manner. *Source* Julian Bolleter

[1]The Australian Bureau of Statistics project, in their Series A projections, that by 2061 Perth's population will have tripled to 6.6 million, Melbourne's doubled to 9.8 million, Sydney's doubled to 8.9 million, and South East Queensland's doubled to 5.6 million (Australian Bureau of Statistics 2017).

livability and viability of Australian cities in this century. As Infrastructure Australia (2018) warns us:

> If we fail to effectively anticipate and respond to growth, the results will be declining economic productivity, increasing environmental pressures and a marked reduction in each city's quality of life.

In such generic statements, we sometimes overlook that real people end up having to live with the consequences of our planning. Real people whose mental health is compromised through the eradication of nature in poorly conceived infill, real people who struggle to access job opportunities in fringe suburbs and are consigned to interminable commutes, real people who find themselves isolated in higher-density TOD settings without the soft infrastructure and public spaces to weave a community and sense of belonging. Of course, GOD does not provide all the answers to the challenges we face, but it could make a substantial contribution.

5.4 Conclusion

Given "urban sprawl" currently outpaces population growth, clearly the lure of suburbia remains for substantial populations around the world. If planners are to effectively deal with the problems of sprawl, and to deliver equitable, efficient and sustainable use of land and natural resources through urban infill development, they need to deliver urban densification in a manner that resonates with the leafy green qualities of suburbia desired by most people (at very least in Australia). The presumption in TOD ideology that future residents will trade the benefits of nature for the benefits of urbanity is a misconception that their proponents promulgate to their own detriment. What we require is a new dream for urban densification, befitting the urban, societal and ecological challenges of the twenty-first century, and aligned with the ongoing preference for suburban living near nature.

As this book attests, we believe that GOD and its suggested well-designed higher-density urban precincts surrounding good quality public green spaces offer a viable, sensible approach to tackle sprawl in Australian and other cities worldwide with similar urbanization patterns.

GOD help us all …

References

Australian Bureau of Statistics (2017) Australian Bureau of Statistics. http://www.abs.gov.au. Accessed 17 Oct 2017

Bolleter J, Weller R (2013) Made in Australia: the future of Australian cities. University of Western Australia Publishing, Perth

Bruegmann R (2017) The anti-suburban crusade. In: Berger A, Kotkin J, Guzman C (eds) Infinite suburbia. MIT, Boston, pp 26–37

Burton P (2017) South East Queensland: change and continuity in planning. In: Hamnett S, Freestone R (eds) Planning metropolitan Australia. Routledge, pp 156–177

Dovey K, Woodcock I (2014) Intensifying Melbourne: transit-oriented urban design for resilient urban futures. Melbourne School of Design, The University of Melbourne, Melbourne

Gleeson B, Dodson J, Spiller M (2010) Metropolitan governance for the Australian city: the case for reform. Issues Pap 12(1):1–26

Goodman R (2017) Melbourne: growing pains for the liveable city. In: Hamnett S, Freestone R (eds) Planning metropolitan Australia. Routledge, pp 59–83

Goodman R, Moloney S (2004) Activity centre planning in Melbourne revisited. Aust Plan 41(2)

Hurley J, Taylor E, Dodson J (2017) Why has urban consolidation been so difficult. In: Sipe N, Vella K (eds) The Routledge handbook of Australian urban and regional planning, pp 123–135

Infrastructure Australia (2018) Future cities: planning for our growing population (summary report). Australian Government, Canberra

Kelly J-F, Donegan P (2015) City limits: why Australian cities are broken and how we can fix them. Melbourne University Press, Melbourne

Randolph B, Freestone R, Bunker R (2017) Sydney: growth, globalization and governance. In: Hamnett S, Freestone R (eds) Planning metropolitan Australia. Routledge, pp 84–108

Troy P (2004) Saving our cities with suburbs. In: Schultz J (ed) Griffith review: dreams of land. Griffith University, Brisbane

MIX
Papier aus verantwortungsvollen Quellen
Paper from responsible sources
FSC® C105338

If you have any concerns about our products,
you can contact us on
ProductSafety@springernature.com

In case Publisher is established outside the EU,
the EU authorized representative is:
Springer Nature Customer Service Center GmbH
Europaplatz 3, 69115 Heidelberg, Germany

Printed by Libri Plureos GmbH
in Hamburg, Germany